Monographs on
Theoretical and Applied Genetics 23

Monographs on Theoretical and Applied Genetics

Volume 10 **Male Sterility in Higher Plants**
By M.L.H. Kaul (1988)

Volume 11 **Tree Breeding: Principles and Strategies**
By G. Namkoong, H.C. Kang, and J.S. Brouard (1988)

Volume 12 **The Wheat Rusts – Breeding for Resistance**
By D.R. Knott (1989)

Volume 13 **Population Dynamics and the Tribolium Model Genetics and Demography**
By R.F. Costantino and R.A. Desharnais (1991)

Volume 14 **Genetic Improvement of Tomato**
Editor: G. Kalloo (1991)

Volume 15 **Evolution and Function of Heterostyly**
Editor: S.C.H. Barrett (1992)

Volume 16 **Distant Hybridization of Crop Plants**
Editors: G. Kalloo and J.B. Chowdhury (1992)

Volume 17 **Cytogenetics in Genetics and Plant Breeding**
By J. Sybenga (1992)

Volume 18 **Cytogenetics of the Festuca – Lolium Complex**
By P.P. Jauhar (1993)

Volume 19 **Breeding Oilseed Brassicas**
Editors: K.S. Labana, S.S. Banga, and S.K. Banga (1993)

Volume 20 **Oenothera**
By C. Harte (1994)

Volume 21 **Soil Mineral Stresses**
Editors: A.R. Yeo and T.J. Flowers (1994)

Volume 22 **Heterosis and Hybrid Rice Breeding**
By S.S. Virmani (1994)

Volume 23 **Biotechnology in Forage and Turf Grass Improvement**
By G. Spangenberg, Z. Wang and I. Potrykus (1998)

G. Spangenberg Z.-Y. Wang I. Potrykus

Biotechnology in Forage and Turf Grass Improvement

With 32 Figures, 16 in Colour

Springer

Cooperative Research Centre Molecular Plant Breeding

Prof. Dr. GERMÁN SPANGENBERG
Dr. ZENG-YU WANG

Plant Sciences and Biotechnology
Agriculture Victoria
Department of Natural Resources and Environment
and CRC for Molecular Plant Breeding
La Trobe University
3083 Bundoora, Victoria
Australia

Prof. Dr. INGO POTRYKUS
Institute of Plant Sciences
Swiss Federal Institute of Technology
8092 Zürich, Switzerland

ISBN 978-3-642-72053-6 **ISBN 978-3-642-72051-2** (eBook)
DOI 10.1007/978-3-642-72051-2

Library of Congress Cataloging-in-Publication Data
Spangenberg, G. (German), 1959–
Biotechnology in forage and turf grass improvement / G. Spangenberg, Z. Wang, I. Potrykus.
p. cm. — (Monographs on theoretical and applied genetics : 23)
Includes bibliographical references and index.

1. Forage plants—Biotechnology. 2. Turfgrasses—Biotechnology.
I. Wang, Z., 1963– . II. Potrykus, I. (Ingo), 1933–
III. Title. IV. Series.
SB193.S668 1998
633.2—dc21

Softcover reprint of the hardcover 1st edition 1998

Cover design: Design & Production GmbH, Heidelberg

Typesetting: Best-Set Typesetter Ltd., Hong Kong

SPIN: 10521189 31/3137 – 5 4 3 2 1 0 – Printed on acid-free paper

Acknowledgements

We are very grateful to our colleagues and friends Sue Dalton, Gilles Charmet, Kristina Glimelius, Robyn Heath, Jillian Hinch, Gaston Legris, Hans-Ferdinand Linskens, Annette Olesen, Rafael Perez-Vicente, Ulrich Posselt, Barry Scott, Howard M Thomas, Xinli Wu and Xudong Ye for reading the various chapters, for providing comments and for contributing artwork.

Our thanks also to Gabi, Kui-hua and Inge for their patience and permanent support.

Contents

Chapter 1
Introduction

Worldwide, acreage under grassland is estimated to be twice that of cropland. Two closely related genera, *Festuca* L. (fescues) and *Lolium* L. (ryegrasses) are of significant value in temperate grasslands. These genera (tribe Poeae, subfamily Pooideae) contain well-adapted, very productive grasses widely distributed in temperate and cool climates in Europe, North and South America, North, East and South Africa, Asia, Australia and New Zealand, where they are used for agricultural and recreational purposes (Jauhar 1993). They are important for grazing, stabilizing soil for agriculture, and enhancing the environment through multiple uses, such as forage, conservation and turf (Barnes 1990). Therefore, in the family Poaceae, the *Festuca-Lolium* group of grasses is among the most extensively studied by agronomists, plant breeders, animal scientists, taxonomists and cytogeneticists. The potential of biotechnological approaches has been recognized for the development of improved fescue and ryegrass cultivars (Barnes 1990; Kaul 1990; Jauhar 1993).

1.1 Agronomic Importance of the *Festuca-Lolium* Complex

The *Festuca-Lolium* complex involves some well-adapted, highly productive persistent species which are widely used for soil stabilization, for agricultural purposes and as amenity grasses. For temperate grasslands, tall fescue, meadow fescue, Italian ryegrass and perennial ryegrass are particularly important species which show complementary desirable traits, such as palatability and fast initial growth of the ryegrasses, and winter hardiness, persistency combined with continued high production after the second harvest year of the fescues. These grasses form the foundation of grassland agriculture and thus represent an important resource contributing to the world economy (Jauhar 1993).

1.1.1 Major *Festuca* Species

Festuca L. is a large, widely distributed and diverse genus comprising approximately 450 perennial species (Clayton and Renvoize 1986). In the fescues, different ploidy levels ranging from diploid (2n = 2x = 14) to decaploid (2n = 10x = 70) are found (Borrill et al. 1971). Fescues of significant agricultural importance are the allohexaploid tall fescue (*F. arundinacea* Schreb.; 2n = 6x = 42) and the diploid meadow fescue (*F. pratensis* Huds.; 2n = 2x = 14), which belong to the Bovinae section (broad-leaved fescues). Other hexaploid fescues of relevance are red or golf course fescue (*F. rubra* L.) and sheep fescue (*F. ovina* L.); both are members of the section Ovinae (fine-leaved fescues; Jauhar 1993).

F. arundinacea is a wind-pollinated, highly self-infertile polyploid perennial cool-season forage and turfgrass which serves as perennial ground cover for millions of acres of erodible land (Barnes 1990). Tall fescue has become the predominant cool-season perennial grass species in the US where it is grown on approximately 35 million acres (Buckner et al. 1979). Some of this acreage has resulted from natural seeding, but much of it is due to introduced seeding (Barnes 1990). Tall fescue provides forage to millions of sheep and cattle in different grassland countries (Buckner et al. 1979). Although tall fescue has many good qualities (e.g. vigor, yield, hardiness, persistency) as a pasture grass, it lacks palatability and is low in nutritional value (Asay et al. 1979).Tall fescue is also used in lawns, parks, golf courses, football fields, highway medians and roadsides. The US seed production of turf-type tall fescue is estimated at 20 million kg yearly (Meyer and Funk 1989).

F. pratensis is a major cool-season, high-yielding, nutritious forage grass considered to be the most important diploid species in terms of the extent of geographical distribution. Since meadow fescue has good digestibility, good winter hardiness, good late summer production, and longevity under a system with frequent cutting or grazing, it has become an increasingly important crop species in leys and pastures (Aastveit and Aastveit 1989; Jauhar 1993). It represents the most important diploid *Festuca* species in the European Union (EU) with 4 million kg of meadow fescue seed required annually (Kley 1994).

F. rubra is a fine-leaved persistent turfgrass well suited for fine high-use lawns and golf courses, but also considered to be important as a forage species. In the EU, the seed consumption of red fescue is estimated at 17 million kg yearly (Kley 1994). It is a variable species which is important both ecologically as a major constituent of North-temperate grassland and economically as a contributor to turfgrass seed production in Britain where it is second to *Lolium* in terms of volume (Zaghmout and Torello 1990; Jauhar 1993).

1.1.2 Major *Lolium* Species

Lolium L. is a small genus which includes only eight diploid (2n = 2x = 14) species (Terell 1966). The commercially most important ryegrasses in cool temperate climates throughout the world are Italian or annual ryegrass (*L.*

multiflorum Lam.) and perennial ryegrass (*L. perenne* L.). In addition, the corresponding fully fertile interspecific hybrid ryegrass (*L.* × *boucheanum* Kunth.) shows complementary agronomic features from both perennial and Italian ryegrasses. In Europe these outcrossing species are particularly important as forage grasses, representing approximately 70% of the grass seed volume required (Kley 1994). They are the key forage species in the Netherlands, UK and New Zealand – countries where livestock production is an intensive enterprise (Buckner et al. 1967).

L. multiflorum is a biennial (var. *italicum* Beck) or annual (var. *westerwoldicum* Mansh.) highly palatable nutritious grass which shows a rapid establishment from seed, good production in the seeding year and rapid recovery after defoliation (Buckner et al. 1967; Jauhar 1993). Italian and Westerwold ryegrasses represent – with an estimated 36 million kg of seed – 27% of the yearly grass seed consumption in the EU (Kley 1994).

L. perenne is a palatable, persistent grass of high tillering density that shows resistance to treading and good response to high nitrogen application. It was probably the first herbage grass to become a crop plant (Borrill 1976). In New Zealand, perennial ryegrass is grown on 7 million ha providing high quality forage to support 60 million sheep and cattle (Siegel et al. 1985). Perennial ryegrass is – with an estimated 55 million kg of seed consumed annually – the main grass species in the EU (Kley 1994). In the US, perennial ryegrass is grown in permanent pastures, particularly on the Pacific Coast and in the southern states. It is also a common component of lawn grass mixtures, with an estimated 28 million kg of turf-type perennial ryegrass seed planted annually in the US (Meyer and Funk 1989).

1.2 Distribution of Fescues and Ryegrasses

Fescues and ryegrasses are typically distributed in the temperate zone and are largely adapted to a variety of eco-geographical habitats. They originated in Europe and on the borders of Asia, and have a wide natural distribution which has been extended even further by man (Jauhar 1993).

The hexaploid *F. arundinacea* and the diploid *F. pratensis* are dominant, widespread fescues in eastern and central Europe, Scandinavia and the northern Mediterranean. Tall fescue is indigenous to Europe and also naturally occurs on the Baltic coast, throughout the Caucasus and in western Siberia, extending into China. Introductions have been made into North and South America, Australia, New Zealand, Japan, and South and East Asia (Barnes 1990).

Ryegrasses are indigenous to temperate regions of Asia and Europe, the North Atlantic Islands and North Africa. They were introduced by man in the Americas, Australia and New Zealand (Jauhar 1993).

1.3 Biotechnology in *Festuca-Lolium* Improvement: General Considerations

Genetic improvement of tall, meadow and red fescues by conventional plant breeding is very slow since these *Festuca* species are predominantly, if not completely, allogamous wind-pollinated grasses and generally self-sterile (Kasperbauer 1990; Kaul 1990; Jauhar 1993). This is also true for the outbreeding ryegrasses *L. multiflorum* and *L. perenne* (Jauhar 1993). These features represent a great challenge to the breeder because these fescue and ryegrass plants are usually heterozygous, and self-incompatibility limits inbreeding to concentrate desired genes for use in rapid development of new cultivars (Kaul 1990). Although many improved forage- and turf-type grass varieties have been developed through recent breeding efforts, progress in improving particular traits has slowed (Ha et al. 1992). Therefore, biotechnological approaches such as anther culture for the generation of haploids, somatic hybridization for combining germplasm from sexually incompatible species and for transferring organellar-coded traits, genetic transformation for the direct introduction of agronomically useful genes, and use of molecular markers as tools in cultivar identification and marker-assisted selection show promise when considered as part of *Festuca-Lolium* improvement programs (Gray 1990; Kaul 1990; Zaghmout and Torello 1990; Ha et al. 1992; Jauhar 1993; De Loose et al. 1993). These approaches are expected to complement, accelerate or enhance conventional breeding in fescues and ryegrasses, since they lead to a broader range of sources from which genetic information may be obtained for improvement and allow the study of breeding strategies as well as offer ways to identify varieties, protect breeder rights and determine seed purity.

References

Aastveit AH, Aastveit K (1989) Genetic variations and inheritance of quantitative characters in two populations of meadow fescue (*Festuca pratensis* Huds.) and their hybrid. Hereditas 111: 103–114

Asay KH, Frakes RV, Buckner RC (1979) Breeding and cultivars. In: Buckner RC, Bush LP (eds) Tall fescue. American Society of Agronomy, Madison, pp 111–139

Barnes RF (1990) Importance and problems of tall fescue. In: Kasperbauer MJ (ed) Biotechnology in tall fescue improvement, Chap 1. CRC Press, Boca Raton, pp 2–12

Borrill M (1976) Temperate grasses. *Lolium, Festuca, Dactylis, Phleum, Bromus* (Gramineae). In: Simmonds NW (ed) Evolution of crop plants. Longman, London, pp 137–142

Borrill M, Tyler B, Lloyd-Jones M (1971) Studies in *Festuca*. I. A chromosome atlas of Bovinae and Scariosae. Cytologia 36: 1–14

Buckner RC, Todd JR, Burrus PB, Barnes RF (1967) Chemical composition, palatability, and digestibility of ryegrass-tall fescue hybrids, "Kenwell", and "Kentucky 31" tall fescue varieties. Agron J 59: 345–349

Buckner RC, Powell JB, Frakes RV (1979) Historical development. In: Buckner RC, Bush LP (eds) Tall fescue. American Society of Agronomy, Madison, pp 1–8

Clayton WD, Renvoize SA (1986) Genera *Graminum*. Grasses of the world. Kew bulletin additional series XIII. Her Majesty's Stationery Office, London, pp 87–109
De Loose M, Vauterin L, Reuheul D, Van Bockstaele E (1993) Identification of rye-grass varieties using RAPD markers. Med Fac Landbow Univ Gent 58/4b: 2119–2121
Gray DJ (1990) Somatic cell culture and embryogenesis in Poaceae. In: Kasperbauer MJ (ed) Biotechnology in tall fescue improvement. CRC Press, Boca Raton, pp 25–57
Ha SB, Wu FS, Thorne TK (1992) Transgenic turf-type tall fescue (*Festuca arundinacea* Schreb.) plants regenerated from protoplasts. Plant Cell Rep 11: 601–604
Jauhar PP (1993) Cytogenetics of the *Festuca-Lolium* complex. Relevance to breeding. In: Frankel R, Grossman M, Linskens HF, Maliga P, Riley R (eds) Monographs on theoretical and applied genetics, vol 18. Springer, Berlin Heidelberg New York, 243 pp
Kasperbauer MJ (1990) Plant regeneration and evaluation. In: Kasperbauer MJ (ed) Biotechnology in tall fescue improvement. CRC Press, Boca Raton, pp 59–77
Kaul K (1990) Potential biotechnological approaches. In: Kasperbauer MJ (ed) Biotechnology in tall fescue improvement. CRC Press, Boca Raton, pp 13–23
Kley G (1994) Breeding and use of forage plants in Europe. In: Krohn K, Paul VH, Thomas J (eds) International conference on harmful and beneficial microorganisms in grassland, pastures and turf, vol 17. IOBC/West palaearctic regional section bulletin, Wageningen, pp 9–12
Meyer WA, Funk CR (1989) Progress and benefits to humanity from breeding cool-season grasses for turf. In: Sleper DA, Asay KH, Pedersen JF (eds) Contributions from breeding forage and turf grasses. CSSA special publication 15, Crop Science Society of America, Madison, pp 31–48
Siegel MR, Latch GCM, Johnson MC (1985) *Acremonium* fungal endophytes of tall fescue and perennial ryegrass: significance and control. Plant Dis 69: 179–183
Terrell EE (1966) Taxonomic implications of genetics in ryegrasses (*Lolium*). Bot Rev 32: 138–164
Zaghmout OMF, Torello WA (1990) Isolation and culture of protoplasts from embryogenic suspension cultures of red fescue (*Festuca rubra* L.). Plant Cell Rep 9: 340–343

Chapter 2
Meristem Culture

2.1 Introduction

Shoot meristems, whether apical or axillary, are small domes (50–150 μm in diameter) consisting of groups of a few actively dividing cells protected by developing leaves. The main physiological functions of meristems are the synthesis of protoplasm and the production of new cells. The biology and developmental morphology of the shoot apical meristems have been studied in grasses (Evans and Grover 1940; Sharman 1942, 1945, 1947; Peacock 1975; Parsons and Robson 1980; Sweet et al. 1991).

In general, meristem culture involves removal and in vitro culturing of the domes of meristematic cells from main shoots and/or axillary buds. However, besides "meristem culture" a striking disparity of terminology is apparent in the literature, and terms such as "shoot tip culture", "meristem tip culture", "culture of shoot apices" or "tip culture" are also often used, particularly due to the variable explant sizes used by different authors. The cultured explants frequently consist of the meristematic dome with one or more leaf primordia to increase chances for further development (Quak 1977; Carman 1995).

Cultured shoot apical meristems can synthesize the required biomolecules for growth and development from a simple, auxin-containing, chemically defined nutrient medium, such as MS (Murashige and Skoog 1962), and differentiate into plantlets in vitro (Smith and Murashige 1970). However, perturbed meristem development in vitro has been observed, and more complete nutrient regimes that mimic those found in situ have been proposed to ameliorate these developmental anomalies (Carman 1995).

Meristem cultures mainly provide a means of vegetative propagation, in vitro germplasm storage and virus eradication. They have therefore found extensive application – particularly for vegetatively propagated horticultural crops – for over 40 years (Murashige 1974; Quak 1977; Hussey 1978; Kartha 1986; Van Zaayen et al. 1992). Meristem culture has received additional interest as a tool in plant transformation studies (Sautter 1993).

Meristem culture for micropropagation and in vitro germplasm storage has been described in forage grasses, although in this context it is undoubtedly not as commercially compelling as in normally vegetatively propagated crops (Table 2.1). In forage grasses, this tissue culture technique obviously has

Table 2.1. Meristem culture in forage and turf grasses

Plant species	Explant	Culture response	References
Festuca arundinacea (tall fescue)	Meristem tip	Plants	Dale (1977b)
	Immature inflorescence	Plants	Dale and Dalton (1983)
Festuca pratensis (meadow fescue)	Meristem tip	Plants	Dale (1977b)
	Immature inflorescence	Plants	Dale and Dalton (1983)
Festuca rubra (red fescue)	Meristem tip	Plants	Dale (1977b)
Lolium perenne (perennial ryegrass)	Meristem tip	Plants	Dale (1977b)
	Meristem tip	Plantlets	Humphreys and Dalton (1992)
	Immature inflorescence	Plants	Dale and Dalton (1983)
	Meristem tip	Fertile plants	Pérez-Vicente et al. (1993)
Lolium multiflorum (Italian ryegrass)	Meristem tip	Plants	Dale (1975, 1977a, b, 1980)
	Meristem tip	Plantlets	Dalton and Dale (1981, 1985)
	Meristem tip	Plantlets	Dale and Webb (1985)
	Meristem tip	Fertile plants	Pérez-Vicente et al. (1993)
	Immature inflorescence	Plantlets	Dale et al. (1981)
	Immature inflorescence	Plants	Dale and Dalton (1983)
	Immature inflorescence	Viable pollen	Pérez-Vicente et al. (1993)
Lolium × *boucheanum* (hybrid ryegrass)	Meristem tip	Plants	Dale (1977b)
	Immature inflorescence	Plants	Dale and Dalton (1983)
Lolium multiflorum × *Festuca pratensis* (*Festulolium* hybrid)	Immature inflorescence	Plants	Dale and Dalton (1983)
Dactylis glomerata (cocksfoot)	Meristem tip	Plants	Dale (1977b, 1979)
	Immature inflorescence	Plants	Dale and Dalton (1983)
Phleum pratense (timothy)	Meristem tip	Plants	Dale (1977b)
	Immature inflorescence	Plants	Dale and Dalton (1983)

potential value for breeding and production of novel cultivars as well as for research programs. The maintenance of (1) specific heterozygous cultivar mother plants from which seeds need to be multiplied over several generations, (2) sexually sterile plants such as haploids or sterile mutants, and (3) rare aneuploids or plants with unusual and meiotically unstable chromosomal compositions illustrates some potential application areas. Severe problems of maintenance – particularly with short-lived grasses – as well as problems of disease accumulation encountered when these plants are conventionally field- or greenhouse-propagated could be overcome in this way (Dale and Webb 1985).

Reports on culture and precocious development of vegetative shoot apices, incompletely determined floral primordia, floral apices and immature inflorescences in forage grasses with emphasis on fescues and ryegrasses are reviewed below (Table 2.1). Studies using these explants for the establishment of callus cultures followed by plant regeneration are referred to in Chapter 3.

2.2 Culture of Vegetative Meristems in *Festuca* and *Lolium*

For grasses belonging to the *Festuca-Lolium* complex, a meristem tip culture technique allowing regeneration of complete plants was first reported by Dale (1975) for the eradication of viruses in *L. multiflorum*. In this study, out of three basal culture media tested, the Murashige and Skoog (MS) medium gave the best results concerning frequency of plant regeneration. Supplementing the MS basal medium with 0.2 mg/l kinetin favoured normal in vitro development of the 0.2- to 1.5-mm long excised vegetative shoot apices of Italian ryegrass, and plant regeneration frequencies of up to 56% were observed. A marked effect of the meristem tip size on survival after 5 weeks of in vitro culture was found, with 76% survival for cultured meristems between 0.2 and 0.5 mm long, which increased to 90% for shoot tips between 0.5 and 1.5 mm long (Dale 1975).

Using this technique, ryegrass mosaic virus (RMV) was eliminated from infected Italian ryegrass plants by culturing meristem tips between 0.2 and 1.1 mm long, which thus consisted of the meristem dome plus one or several pairs of leaf primordia (Dale 1977a). For the elimination of RMV, no advantage in using 2,4-dichlorophenoxyacetic acid (2,4-D) in the culture medium or in culturing small meristems within the 0.2- to 1.1-mm size range was found. Culture of the larger meristem tips for RMV eradication in *L. multiflorum* seems advisable because of their comparative ease of dissection, their higher rate of survival in culture and the better, more rapid growth of corresponding regenerants upon their transfer to soil (Dale 1977a).

This meristem tip culture method originally developed for Italian ryegrass (Dale 1975, 1977a) also proved successful for the culture of shoot apices in different *Lolium* and *Festuca* species (Dale 1977b). A basal MS medium supplemented with 0.01 mg/l 2,4-D and 0.2 m/l kinetin was found to be the best general medium for these species and gave plant regeneration rates of 8% for *F. rubra*, 20% for *L. perenne*, 29% for *F. pratensis* and *F. arundinacea*, 53% for *L.* × *boucheanum*, 53% for tetraploid *L. multiflorum* and 92% for diploid *L. multiflorum* (Dale 1977b). However, the variations in survival rates of in vitro-cultured meristem tips observed between cultivars and species correlated well with the meristem tip size at culturing – the longer the tip, the higher the survival rate – and thus cannot be interpreted as due to a more fundamental difference between cultivars and species. On average, 96% of the regenerated plants could be established in soil (Dale 1977b).

Pérez-Vicente et al. (1993) have reported the in vitro culture of shoot apical meristems excised as 0.05- to 0.3-mm explants, which are potential targets for microballistic transformation in ryegrasses, as discussed in Section 8.3. This meristem culture system allowed for the efficient regeneration of fully fertile plants in *L. multiflorum* and *L. perenne* (Fig. 2.1A–D). Over 80% of the explanted vegetative apices differentiated first leaves within 2 weeks in culture (Fig. 2.1A, B). On average, 79 and 99% of the vegetative meristems of Italian and perennial ryegrass cultured on solidified MS-based medium led to rooted in vitro-growing plantlets consisting of a single tiller, respectively. Over 80% of the soil-grown ryegrass plants derived from the cultured vegetative meristems flowered (Fig. 2.1C). Corresponding anthers produced up to 90% viable pollen, crosses made in both directions succeeded and plants set 92% germinable seeds (Fig. 2.1D; Pérez-Vicente et al. 1993).

These methods for culture of vegetative apices in forage grasses to eradicate certain viruses may find application in plant breeding and seed production to maintain – alive and healthy for many years – desirable plant genotypes, e.g. the basic plants of cultivars used as parent material for seed multiplication prior to commercial release (Dale 1977a, b). Meristem culture could then be a tool for the in vitro storage of superior germplasm in forage grasses. This potential has been demonstrated, particularly for Italian ryegrass (Dale 1980).

Fig. 2.1A–J. Culture of vegetative meristems and immature inflorescences of *Lolium multiflorum* Lam. **A** Isolated vegetative shoot apex with meristematic dome of *L. multiflorum* cv. Axis. Arrow shows line of excision in the final explant; *bar* represents 100 μm. **B** Isolated vegetative apex after 4 days in culture. Oldest leaf primordium from explant has grown and covered the apical meristem; *bar* represents 200 μm. **C** Flowering plants from cultured vegetative apices. **D** Germinated seeds recovered after crossing plants regenerated from cultured vegetative apices. **E** Isolated early immature inflorescence of *L. multiflorum* cv. Axis. Arrow shows line of excision of final explant; *bar* represents 100 μm. **F** Early immature inflorescence as in **E** after 10 days in liquid culture; *bar* represents 500 μm. **G** 20-day-old immature spike developed from an early immature inflorescence in liquid culture. **H** In vitro culture on solidified medium of spikelets explanted from developing spike in **G** which have developed from flower meristems of cultured early immature inflorescences. **I** In vitro-cultured 4-week-old spikelet with protruding stigmas and showing dehisced anthers. **J** Viable pollen from in vitro culture of early immature inflorescences. (Pérez-Vicente et al. 1993)

A
B
C
D
E
F
G
H
I
J

In *L. multiflorum*, after 4–6 weeks of in vitro culture at 25 °C and high light intensity, 2- to 4-cm plantlets grew from the excised 0.3- to 0.9-mm shoot tips and some produced multiple tillers (Dale 1980). Upon transfer of these in vitro-grown plantlets to low light intensity and 2–4 °C – culture conditions for minimal growth, storage for 2–3 years without fresh supply of culture medium was possible (Dale and Webb 1985). However, routine transfer every year to new culture media and vessels, using either shoot tips, tiller bases or small tiller buds as explants, is recommended, since the life-span of plantlets kept under these minimal growth conditions may vary. Subcultured plantlets are allowed to become established in the new culture vessels at 25 °C with high light conditions prior to being returned to storage conditions. In this way specific grass genotypes in *Lolium* and *Festuca* species have been maintained for over 8 years (Dale et al. 1980; Dale and Webb 1985). Mother plants of *L. multiflorum* can grow and set seed normally after 12 years' storage (Dalton 1995, pers. comm.).

In addition to applications in the in vitro storage of aseptic plantlets of forage grasses, meristem culture may be instrumental in clonal propagation (Dale et al. 1980). In this case, aseptic plantlets established from shoot tips or used directly from in vitro storage were trimmed as single tillers which were transferred to propagation medium containing the synthetic cytokinin 6-benzylaminopurine, but no auxin (Dale and Webb 1985). The suppressed dominance of the apical meristems led to rapid production of tillers. In *L. multiflorum* this treatment of a single tiller allowed the production of 40 tillers within 1 month, and the further subcultures of explanted propagules, consisting of about two tillers from the previous micropropagation cycle, resulted again in the generation of 20 propagules each month (Dalton and Dale 1981). In addition, whole tillers grown from shoot tips cultured on regeneration medium were shown to be better suited for propagation on tiller induction medium, since survival of shoot tips on tiller induction medium was low (Dalton and Dale 1981). Further optimization work on in vitro tiller induction in Italian ryegrass showed that in vitro tillers grown on cytokinin-containing media could produce new tillers more quickly (13–60 per 5-week period) than soil-grown tillers (8–15 per 5-week period; Dalton and Dale 1985). Thus in vitro tiller induction could be used in *L. multiflorum* as a method for propagation of cultivar mother plants stored in vitro in order to increase, upon their transfer to soil, the corresponding amount of basic seed produced.

The in vitro tillering from shoot tips has been studied for over 1000 genotypes from nine different cultivars in *L. perenne* (Humphreys and Dalton 1992). These in vitro cultures were assessed for survival, tiller production, size and vigour. The mean number of tillers produced by a single tiller in vitro on corresponding induction medium was much lower for *L. perenne* than values reported for *L. multiflorum* (Dalton and Dale 1985; Humphreys and Dalton 1992). A significant relationship was found between in vitro tiller production and heading date for the cultivars of perennial ryegrass tested, with genotypes of less persistent, early-heading cultivars producing more tillers than genotypes of late-heading, more persistent cultivars (Humphreys and Dalton 1992).

2.3 Culture of Floral Meristems in *Festuca* and *Lolium*

Direct production of shoots from in vitro-cultured immature inflorescences has been reported for fescues and ryegrasses (Dale et al. 1981; Dale and Dalton 1983). Excised 1- to 50-mm long immature inflorescences were induced to form – instead of florets – leaves and shoots or embryoids which germinated to plantlets in *L. multiflorum* (Dale et al. 1981). The frequencies of cultured immature inflorescences undergoing the two different morphogenic pathways varied with the culture media used. Furthermore, the response in culture was also affected by the developmental stage of the explanted immature inflorescences. Direct leaf production was observed in 43% of the cases when 1- to 6-mm immature inflorescences were cultured, and in 22% of the cases for 7- to 50-mm long cultured inflorescences (Dale et al. 1981).

Similar observations on development of shoots instead of florets from floret primordia have been described after in vitro culture of 1- to 20-mm immature inflorescences in *L. perenne*, *L.* × *boucheanum*, *F. arundinacea*, *F. pratensis* and *L. multiflorum* × *F. pratensis* (Dale and Dalton 1983). Shoot production from the in vitro-cultured immature inflorescences was particularly dependent on the developmental stage of the inflorescence at culturing and on the culture medium used. However, other factors, e.g. genotype and conditions of donor plant, might also play a role. Generally, the younger the inflorescence and the less floral-committed the primordia, the greater the number of plantlets produced per cultured immature inflorescence (Dale and Webb 1985). Averaged over species and media used, 5–15 plantlets could be regenerated per cultured immature inflorescence. In addition, high frequencies of plantlet survival in establishment medium (44–100%) and in soil (68–100%) have been reported for different *Festuca* and *Lolium* species (Dale and Dalton 1983). Therefore this method also shows potential for clonal propagation and the establishment of aseptic plantlets for in vitro storage or experimentation in fescues and ryegrasses.

Other interesting applications of these in vitro culture systems for florally induced meristems and immature inflorescences are found in studies on the physiology of flowering in forage grasses. Studies on these lines are available for *L. temulentum*, a strict long day (LD) plant which can be induced to flower when 6-week-old plants are exposed to a single LD (McDaniel et al. 1991; King et al. 1993). Shoot apices of *L. temulentum* excised after exposure of donor plants to 1 LD were found to undergo floral development in vitro on a medium containing gibberellic acid, whereas those explanted from plants kept in short day remained vegetative (McDaniel et al. 1991). This study provided evidence for the timepoint of floral evocation of the shoot apex after exposure to an inductive LD. Conditions allowing for a largely quantitative response to LD induction and normal floral differentiation in vitro by the florally induced and determined shoot apices were identified (McDaniel et al. 1991). Furthermore, the potential of the in vitro culture of excised *Lolium* shoot apices as an assay system for the presence of both florigenic and flower-inhibitory compounds has been demonstrated (King et al. 1993).

A stepwise culture system has been reported for the in vitro development of 0.5- to 1.8-mm floral apices until maturation of floral organs in *L. multiflorum* (Fig. 2.1E–J; Pérez-Vicente et al. 1993). Over 80% of the explanted early immature inflorescences differentiated spikelet primordia (Fig. 2.1F) and produced immature spikes (Fig. 2.1G) with 3–5 well-developed spikelets each. Upon transfer of the excised in vitro-developed spikelets onto solid culture medium under optimal gas exchange conditions (Fig. 2.1H), most of them were found to reach anthesis (Fig. 2.1I). Corresponding anthers showed 44% viable pollen (Fig. 2.1J) which could be germinated in vitro. Crosses made by pollinating emasculated florets of seed-grown plants of *L. multiflorum* with the in vitro-matured pollen from the cultured early immature inflorescences succeeded in 10% of the cases (Pérez-Vicente et al. 1993). Potential applications of this culture system can be foreseen in studies on genetic transformation as well as on differentiation of floral organs and effects of bioregulators, e.g. gametocides and apomictic factors, possibly in combination with direct manipulation via microtargeting (see Sect. 8.3).

2.4 Meristem Culture in Other Grasses

Culture of vegetative and floral apices has been reported for different cereals and grasses (Dale 1977b, 1979; Raman et al. 1980; Dale and Dalton 1983; Dale and Webb 1985; Pareddy et al. 1987; Trione and Stockwell 1989; Stapelton and Bedinger 1992).

In forage and turf grasses other than fescues and ryegrasses, information on meristem culture for virus eradication, in vitro storage and micropropagation is available for cocksfoot or orchardgrass (*Dactylis glomerata*) and timothy (*Phleum pratense*; Dale 1977b, 1979; Dale and Dalton 1983). Up to 31 and 50% of in vitro-cultured vegetative shoot apices have been shown to regenerate to plants in *P. pratense* and *D. glomerata*, respectively (Dale 1977b). Meristem tip culture thus opened opportunities for in vitro storage of aseptic plantlets of specific genotypes of these grass species over many years (Dale and Webb 1985).

The method has also proven valuable for the elimination of viruses from infected stocks of *D. glomerata* (Dale 1979). Shoot tips and tiller buds 0.2- to 5.1-mm long excised from *D. glomerata* plants infected with combinations of cocksfoot streak virus (CFSV), cocksfoot mild mosaic virus (CMMV) and cocksfoot mottle virus (CFMV) were cultured in vitro. None of the viruses were detectable by electron microscopy or serology in plants regenerated from in vitro-cultured meristems of less than 1.2 mm in size, but the explant size capable of regenerating virus-eradicated healthy cocksfoot plants was found to depend upon the infecting virus (Dale 1979).

The production of shoots and plantlets, either directly from floret primordia or via callus, has also been reported for in vitro-cultured 1- to 20-mm immature inflorescences in *Phleum* and *Dactylis* (Dale and Dalton 1983). On average, 11 and 25 plantlets regenerated per each cultured immature inflorescence in timothy and cocksfoot, respectively. In addition to green plants recovered from the cultured immature inflorescences, some albino plantlets were obtained. The percentage of albino plantlets was 7.8% in *P. pratense* and 1.4% in *D. glomerata*, and thus lower than corresponding values reported for the ryegrasses and meadow fescue (Dale and Dalton 1983). Observations of the regenerated plants over 2 years revealed no gross morphological abnormalities, and chromosome counts were normal in all cases tested. This supports the view that in vitro-cultured plantlets obtained in this way may be useful for storage or propagation of specific genotypes in different forage and turf grasses.

2.5 Summary and Conclusions

Efficient methods for the in vitro culture of vegetative and floral apices have been developed for different forage and turf grasses, primarily in *Lolium* species. They show potential for in vitro germplasm storage, virus eradication and micropropagation. The in vitro culture of shoot tips and immature inflorescences until regeneration of plantlets involves no or little callus formation. Thus a minimal risk of genetic instability during storage and micropropagation would be expected. The available experimental evidence confirms the lack of extensive genetic or cytological variation occurring in plants regenerated from these cultures. On a routine commercial scale, germplasm storage and plant propagation in forage and turf grasses are by seeds. However, for certain purposes maintenance over several years and vegetative propagation of healthy genotypes are needed. In this context, effective virus elimination from infected plants, the limited scope for newly induced genetic variation, and accelerated tiller production by the precocious development of shoots from axillary buds make the in vitro culture of vegetative and floral apices a valid methodical option.

References

Carman JG (1995) Nutrient absorption and the development and genetic stability of cultured meristems. In: Terzi M, Cella R, Falavigna A (eds) Current issues in plant molecular and cellular biology. Kluwer Academic, Dordrecht, pp 393–403

Dale PJ (1975) Meristem tip culture in *Lolium multiflorum*. J Exp Bot 26: 731–736
Dale PJ (1977a) The elimination of ryegrass mosaic virus from *Lolium multiflorum* by meristem tip culture. Ann Appl Biol 85: 93–96
Dale PJ (1977b) Meristem tip culture in *Lolium*, *Festuca*, *Phleum* and *Dactylis*. Plant Sci Lett 9: 333–338
Dale PJ (1979) The elimination of cocksfoot streak virus, cocksfoot mild mosaic virus and cocksfoot mottle virus from *Dactylis glomerata* by shoot tip and tiller bud culture. Ann Appl Biol 93: 285–288
Dale PJ (1980) A method for in vitro storage of *Lolium multiflorum* Lam. Ann Bot 45: 497–502
Dale PJ, Dalton SJ (1983) Immature inflorescence culture in *Lolium*, *Festuca*, *Phleum* and *Dactylis*. Z Pflanzenphysiol 111: 39–45
Dale PJ, Webb KJ (1985) Germplasm storage and micropropagation. In: Bright SWJ, Jones MGK (eds) Cereal tissue and cell culture. Nijhoff/Junk, Dordrecht, pp 79–96
Dale PJ, Cheyne VA, Dalton SJ (1980) Pathogen elimination and in vitro plant storage in forage grasses and legumes. In: Ingram DS, Helgeson JP (eds) Tissue culture methods for plant pathologists. Blackwell Scientific, Oxford, pp 119–124
Dale PJ, Thomas E, Brettell RIS, Wernicke W (1981) Embryogenesis from cultured immature inflorescences and nodes of *Lolium multiflorum*. Plant Cell Tissue Organ Cult 1: 47–55
Dalton SJ, Dale PJ (1981) Induced tillering of *Lolium multiflorum* in vitro. Plant Cell Tissue Organ Cult 1: 57–64
Dalton SJ, Dale PJ (1985) The application of in vitro tiller induction in *Lolium multiflorum*. Euphytica 34: 897–904
Evans MW, Grover FO (1940) Developmental morphology of the growing point of the shoot and the inflorescence in grasses. J Agric Res 61: 481–520
Humphreys NC, Dalton SJ (1992) A relationship between heading date and in vitro tillering propensity in perennial ryegrass (*Lolium perenne* L.). Euphytica 58: 253–257
Hussey G (1978) The application of tissue culture to the vegetative propagation of plants. Sci Progs 65: 185–208
Kartha KK (1986) Production and indexing of disease-free plalnts. In: Withers LA, Alderson PG (eds) Plant tissue culture and its agricultural applications. Butterworths, London, pp 219–238
King RW, Blundell C, Evans LT (1993) The behaviour of shoot apices of *Lolium temulentum* in vitro as the basis of an assay system for florigenic extracts. Aust J Plant Physiol 20: 337–348
McDaniel CN, King RW, Evans LT (1991) Floral determination and in-vitro floral differentiation in isolated shoot apices of *Lolium temulentum* L. Planta 185: 9–16
Murashige T (1974) Plant propagation through tissue cultures. Annu Rev Plant Physiol 25: 135–166
Murashige T, Skoog F (1962) A revised medium for rapid growth and bioassays with tobacco tissue culture. Physiol Plant 15: 473–497
Pareddy DR, Greyson RI, Walden DB (1987) Fertilization and seed production with pollen from in-vitro-cultured maize tassels. Planta 170: 141–143
Parsons AJ, Robson MJ (1980) Seasonal changes in the physiology of S24 perennial ryegrass (*Lolium perenne* L.). 1. Response of leaf extension to temperature during the transition from vegetative to reproductive growth. Ann Bot 46: 435–444
Peacock JM (1975) Temperature and leaf growth in *Lolium perenne*. III. Factors affecting seasonal differences. J Appl Ecology 12: 685–697
Pérez-Vicente R, Wen XD, Wang ZY, Leduc N, Sautter C, Wehrli E, Potrykus I, Spangenberg G (1993) Culture of vegetative and floral meristems in ryegrasses: potential targets for microballistic transformation. J Plant Physiol 142: 610–617
Quak F (1977) Meristem culture and virus-free plants. In: Reinert J, Bajaj YPS (eds) Applied and fundamental aspects of plant cell, tissue, and organ culture. Springer, Berlin Heidelberg New York, pp 598–615
Raman K, Walden DB, Greyson RI (1980) Propagation of *Zea mays* L. by shoot tip culture: a feasibility study. Ann Bot 45: 183–189

Sautter C (1993) Development of a microtargeting device for particle bombardment of plant meristems. Plant Cell Tissue Organ Cult 33: 251–257

Sharman BC (1942) Shoot apex in grasses and cereals. Nature 149: 82–83

Sharman BC (1945) Construction of the shoot apex in cereals and other grasses. Nature 155: 291–292

Sharman BC (1947) The biology and developmental morphology of the shoot apex in Gramineae. New Phytol 46: 20–34

Smith RH, Murashige T (1970) In vitro development of the isolated shoot apical meristem of angiosperms. Am J Bot 57: 562–568

Stapleton AE, Bedinger PA (1992) Immature maize spikelets develop and produce pollen in culture. Plant Cell Rep 11: 248–252

Sweet N, Wiltshire JJJ, Baker CK (1991) A new descriptive scale for early reproductive development in *Lolium perenne* L. Grass Forage Sci 46: 201–206

Trione EJ, Stockwell VO (1989) Development of detached wheat spikelets in culture. Plant Cell Tissue Organ Cult 17: 161–170

Van Zaayen A, Van Eijk C, Versluijs JMA (1992) Production of high quality ornamental crops through meristem culture. Acta Bot Neerl 41: 425–433

Chapter 3

Callus Cultures and Somaclonal Variation

3.1 Introduction

In grasses, somatic cell and tissue cultures in vitro have been initiated from a wide range of explants, such as embryos, roots, leaf bases, immature inflorescences and meristems (Yamada 1977; Vasil 1987). When the surface-sterilized explants are placed on appropriate culture media and maintained axenically under suitable growth conditions, competent cells in the explant may dedifferentiate and divide to produce a proliferating unorganized mass of cells called callus. Initiated calli or calluses may be maintained in vitro by periodic subcultures onto fresh media to establish callus cultures.

The initiation and establishment of in vitro callus cultures in grasses is primarily aimed at the regeneration of plants. Regenerable callus cultures in grasses can be initiated from young, immature tissues at specific stages of development and require for their maintenance potent auxins such as 2,4-dichlorophenoxyacetic acid (2,4-D) in the culture media. Plant regeneration from callus cultures in grasses can occur via organogenesis or somatic embryogenesis.

During organogenesis, single cells or groups of callus cells competent for shoot or root primordia formation will produce meristems and these will finally lead to the differentiation of shoots or roots, respectively. When shoots are differentiated, these can be excised from the callus and upon transfer onto a rooting medium, adventitious roots form to produce in vitro-grown plantlets. Rooted plantlets can be transferred to soil and grown until maturity. During somatic embryogenesis, cells competent for this mode of regeneration will differentiate embryoids bearing a shoot and a root meristem organized simultaneously in a bipolar axis. These somatic embryos may germinate into plants.

Based on their morphology, organization, cellular ultrastructure and capacity to regenerate plants, distinct callus types may be identified. Rooty, shooty, organogenic, embryogenic and non-morphogenic callus types have been described in grass cell cultures (King et al. 1978; Morrish et al. 1987). Callus cultures in grasses may consist of different callus types and thus contain non-morphogenic cells and cells competent for regeneration through somatic embryogenesis and organogenesis. To maintain regenerable callus cultures, organogenic or embryogenic calli need to be identified, selectively enriched in every subculture and generally be maintained on media with relatively high auxin levels. For plant regeneration, the regenerable calli are transferred onto media with reduced or omitted auxin, and frequently with added cytokinins (Gray 1990).

In vitro culture can be stressful on the cultured plant cells and may involve mutagenic processes during callus initiation, callus maintenance and cell differentiation until plant regeneration, leading to variability observed among the cell culture-regenerated plants (Evans 1989; Duncan 1996). This variability, called somaclonal variation may be epigenetic or heritable in nature (Larkin and Scowcroft 1981; Lörz et al. 1988). Thus, plants regenerated from callus cultures in grasses may be genetically uniform or non-uniform. Non-uniform regenerants from callus cultures initiated from a defined somatic explant are called somaclonal variants. The nature of a callus may determine the regeneration ability as well as the genetic stability of the regenerated grass plants. Regeneration via organogenesis has been associated with the recovery of non-uniform plants while the somatic embryogenic pathway has been suggested to be less likely to result in the regeneration of somaclonal variants (Vasil 1983; Morrish et al. 1987). This makes embryogenic callus cultures and embryogenic cell suspensions derived from them (Chap. 5) suitable for clonal propagation purposes and an ideal source of target cells for genetic manipulations (Chaps. 6–8). Organogenic cultures provide a useful source of somaclonal variants derived from both pre-existing genetic variation within the original explants and variation induced during the cell culture phase (Evans et al. 1984; Evans 1989). Additional variation can be generated if organogenic cultures are exposed to chemical mutagens or to radiation (Kaul 1990).

In this chapter plant regeneration from callus cultures is described in detail for species belonging to the *Festuca-Lolium* complex. Developmental morphogenesis and somaclonal variation in cell and tissue cultures of fescues and ryegrasses are discussed. Studies on other forage and turf grasses are referred to in Table 3.1.

3.2 Regeneration from Callus Cultures in *Festuca* and *Lolium*

Although the establishment of axenically grown cell and tissue cultures from forage grasses within the *Festuca-Lolium* complex was first accomplished in the mid 1950s (Norstog 1956), the regeneration of whole plants was not achieved until the late 1970s (Ahloowalia 1975; Kasperbauer et al. 1979; Lowe and Conger 1979).

Norstog (1956) reported in vitro callus growth from cultured endosperm of 9- to 10-day-old seeds from *L. perenne* plated on White's medium (White 1943) supplemented with 20% coconut milk or 0.25% yeast extract. Approximately 20% of the cultured perennial ryegrass endosperms proliferated in

Table 3.1. Callus cultures in forage and turf grasses

Species	Explant	Culture response	References
Festuca arundinacea (tall fescue)	Mature seed and embryo	Calli	Conger et al. (1978)
	Mature embryo	Plants	Lowe and Conger (1979)
	Mature embryo	Calli	Conger et al. (1982)
	Mature embryo	Calli with *Neotyphodium*	Conger and McDaniel (1983)
	Mature embryo	Plants	Reed and Conger (1985)
	Peduncle	*Neotyphodium* infected plants	Johnson et al. (1986)
	Mature embryo	Plants	Eizenga (1989)
	Immature inflorescence	Plants	Eizenga and Dahleen (1990)
	Leaf base	*Neotyphodium* infected plants	Kearney et al. (1991)
	Immature embryo and mature seed	Plants	Takamizo et al. (1994)
	Meristem tip	Fertile plants	Garcia et al. (1994)
Festuca pratensis (meadow fescue)	Mature seed and root	Calli	Atkin and Barton (1973)
	Mature seed	Plants	Wang et al. (1993a)
Festuca rubra (red fescue)	Mature seed and root	Calli	Atkin and Barton (1973)
	Mature seed	Plants	Torello et al. (1984)
	Mature seed	Plants	Torello et al. (1985)
	Mature seed	Plantlets	Zaghmout and Torello (1988, 1992a)
	Mature seed	Plants	Spangenberg et al. (1994)
Lolium multiflorum (Italian ryegrass)	Mature seed and root	Calli	Atkin and Barton (1973)
	Mature seed and embryo	Calli	Conger et al. (1978)
	Immature embryo	Plantlets	Dale (1980)
	Immature inflorescence	Plants	Dale et al. (1981)
	Mature embryo and leaf tips	Calli	Conger et al. (1982)
	Leaf base	Calli	Joarder et al. (1986)
	Root	Plants	Jackson et al. (1986)
	Root, meristem tip and leaf base	Plantlets	Jackson and Dale (1988)
	Immature inflorescence	Plants	Creemers-Molenaar et al. (1988)
	Meristem tip and leaf base	Plants	Jackson and Dale (1989)
	Mature seed	Plants	Wang et al. (1993b)

Table 3.1. *Continued*

Species	Explant	Culture response	References
Lolium perenne (perennial ryegrass)	Endosperm	Calli	Norstog (1956)
	Mature seed and root	Calli	Atkin and Barton (1973)
	Mature seed	Plants	Torello and Symington (1984)
	Immature inflorescence	Plants	Creemers-Molenaar et al. (1988)
	Mature seed and stem segment	Plantlets	Zaghmout and Torello (1992b)
	Mature seed	Plants	Wang et al. (1993b)
	Mature embryo	Fertile plants	Posselt and Altpeter (1994)
	Meristem tip	Plantlets	Olesen et al. (1995)
Lolium temulentum (annual ryegrass)	Mature seed and root	Calli	Atkin and Barton (1973)
	Immature embryo	Fertile plants	Jackson and Dale (1989)
Lolium rigidum (wimmera ryegrass)	Mature seed and immature embryo	Plants	Skene and Barlass (1983)
Lolium multiflorum × *Festuca arundinacea*	Internode and peduncle	Plants	Kasperbauer et al. (1979)
Lolium × *boucheanum*	Triploid immature embryo	Plants	Ahloowalia (1975)
L. multiflorum × *L. perenne* (hybrid ryegrass)	Triploid immature embryo	Plants	Ahloowalia (1983)
	Mature seed	Plants	Wang et al. (1993b)
Agropyron cristatum (crested wheatgrass)	Immature inflorescence	Plants	Lo et al. (1980)
Agropyron repens × *Bromus inermis* (*Agromus*)	Immature inflorescence	Plants	Gyulai et al. (1992)
Agropyron smithii (western wheatgrass)	Immature inflorescence	Plants	Lo et al. (1980)
Agrostis palustris (creeping bentgrass)	Mature seed	Plantlets	Krans et al. (1982)
	Mature seed	Plantlets	Blanche et al. (1986)
	Mature seed	Plants	Zhong et al. (1991)

Agrostis stolonifera	Meristem tip and root	Plants	Wu and Antonovics (1978)
Agrostis tenuis (common bentgrass)	Mature seed and root	Calli	Atkin and Barton (1973)
Alopecurus arundinaceus (creeping foxtail)	Immature inflorescence	Plants	Lo et al. (1980)
Andropogon gerardii (big bluestem)	Immature inflorescence	Plants	Chen et al. (1977)
Bothriochloa caucasica (Caucasian blustem)	Mature seed	Fertile plants	Franklin et al. (1990)
Bothriochloa ischaemum (old world blustem)	Immature inflorescence	Plants	Metzinger et al. (1987)
	Immature inflorescence	Plants	Artunduaga et al. (1988)
Bromus inermis (brome grass)	Immature inflorescence	Plants	Lo et al. (1980)
Cenchrus ciliaris (buffel grass)	Mature seed	Plants	Ross et al. (1995)
Cynodon dactylon (common bermudagrass)	Immature inflorescence	Plants	Ahn et al. (1985, 1987)
	Immature inflorescence	Plants	Artunduaga et al. (1988, 1989)
Cynosurus cristatus (crested dogstail)	Mature seed and root	Calli	Atkin and Barton (1973)
Dactylis glomerata (cocksfoot)	Mature seed and root	Calli	Atkin and Barton (1973)
	Mature seed and embryo	Calli	Conger et al. (1978)
	Mature seed	Plants	Conger and Carabia (1978)
	Mature embryo	Calli	Conger et al. (1982)
	Leaf base	Plantlets	Hanning and Conger (1982)
	Leaf base	Plants	Conger et al. (1983)
	Leaf base	Plants	Hanning and Conger (1986)
	Leaf base	Plantlets	Gavin et al. (1989)

Table 3.1. *Continued*

Species	Explant	Culture response	References
Digitaria sanguinalis	Leaf base and mature seed	Plants	Gonzales and Franks (1987)
Echinochloa colona	Immature inflorescence	Fertile plants	Tyagi et al. (1985)
	Leaf base and mesocotyl	Plants	Samantaray et al. (1995)
	Leaf base and leaf tip	Plants	Samantaray et al. (1997)
Echinochloa crusgalli (barnyard grass)	Immature inflorescence	Plantlets	Wang and Yan (1984)
	Mesocotyl	Plants	Cobb et al. (1985)
Echinochloa frumentacea	Immature inflorescence	Plants	Sankhla et al. (1992)
	Immature inflorescence and embryo	Plants	Talwar and Rashid (1989)
Echinochloa muricata	Mesocotyl	Plants	Cobb et al. (1985)
Echinochloa oryzicola	Leaf sheath	Plants	Takahashi et al. (1984)
Eleusine coracana (finger millet)	Mesocotyl	Plants	Rangan (1976)
	Mesocotyl and leaf base	Plants	Mohanty et al. (1985)
	Mature seed and meristem tip	Plantlets	Eapen and George (1989)
	Mature seed	Plantlets	Eapen and George (1990)
	Mature seed	Fertile plants	Sivadas et al. (1990)
Elymus canadensis (Canada wildrye)	Immature embryo and inflorescence	Plants	Park and Walton (1989ab)
Elymus canadensis × *Psathyrostachys juncea*	Immature inflorescence	Plants	Park et al. (1990)
Elymus canadensis × *Secale cereale*	Immature inflorescence	Plants	Park et al. (1990)
Eragrostis curvula (weeping lovegrass)	Immature inflorescence	Plants	Echenique et al. (1996)

Panicum bisulcatum	Mature seed and mesocotyl	Fertile plants	Fladung and Hesselbach (1986)
Panicum maximum (guinea grass)	Leaf base	Plants	Lu and Vasil (1981)
	Immature inflorescence and embryo	Plants	Lu and Vasil (1982)
	Leaf base	Fertile plants	Hanna et al. (1984)
	Immature embryo	Plants	Akashi and Adachi (1991)
Panicum miliaceum (proso millet)	Mesocotyl	Fertile plants	Rangan (1974)
	Immature embryo, mature seed	Fertile plants	Heyser and Nabors (1982)
	Mesocotyl, leaf and stem segment		
	Immature inflorescence	Fertile plants	Rangan and Vasil (1983)
Panicum miliare (little millet)	Immature inflorescence	Fertile plants	Rangan and Vasil (1983)
Panicum milioides	Leaf base, mesocotyl and seed	Fertile plants	Fladung and Hesselbach (1986)
Panicum virgatum (switchgrass)	Mature seed and leaf base	Plants	Denchev and Conger (1994)
	Mature seed and leaf base	Plants	Denchev and Conger (1995)
Paspalum almum	Immature inflorescence	Plants	Bovo and Mroginski (1986)
Paspalum dilatatum (dallisgrass)	Immature inflorescence	Plants	Akashi and Adachi (1992)
Paspalum dilatatum × *Paspalum regnellii*	Immature inflorescence	Plants	Bovo and Mroginski (1986)
Paspalum equitans	Immature inflorescence	Plants	Bovo and Mroginski (1986)
Paspalum intermedium	Immature inflorescence	Plants	Bovo and Mroginski (1986)
Paspalum notatum (bahiagrass)	Immature inflorescence	Plants	Bovo and Mroginski (1986)
	Mature seed	Plants	Marousky and West (1990)
	Leaf-stem segment	Plants	Shatters et al. (1994)
	Mature seed	Plants	Akashi et al. (1993)
Paspalum ovale	Immature inflorescence	Plants	Bovo and Mroginski (1986)
Paspalum quadrifarium	Immature inflorescence	Plants	Bovo and Mroginski (1986)
Paspalum rufum	Immature inflorescence	Calli	Bovo and Mroginski (1986)

Table 3.1. *Continued*

Species	Explant	Culture response	References
Paspalum scrobiculatum (minor millet)	Mesocotyl	Plants	Rangan (1976)
Paspalum virgatum	Immature inflorescence	Plants	Bovo and Mroginski (1986)
Paspalum wettsteinii	Immature inflorescence	Plants	Bovo and Mroginski (1986)
Pennisetum americanum (pearl millet)	Immature embryo and inflorescence	Plants	Vasil and Vasil (1981)
	Immature embryo	Plants	Vasil and Vasil (1982)
	Immature inflorescence	Fertile plants	Swedlund and Vasil (1985)
	Immature inflorescence	Plantlets	Pius et al. (1993)
Pennisetum americanum × *Pennisetum purpureum*	Immature embryo and inflorescence	Plants	Vasil and Vasil (1981)
	Immature inflorescence	Plants	Rajasekaran et al. (1986)
Pennisetum glaucum (pearl millet)	Immature inflorescence	Fertile plants	Morrish et al. (1990)
	Immature inflorescence	Fertile plants	Nagarathna et al. (1993)
Pennisetum purpureum (napier grass)	Leaf base	Plants	Haydu and Vasil (1981)
	Immature inflorescence	Plants	Wang and Vasil (1982)
	Leaf base	Plantlets	Chandler and Vasil (1984)
	Leaf base and inflorescence	Plants	Shenoy and Vasil (1992)
Pennisetum typhoideum (bulrush millet)	Mesocotyl	Plants	Rangan (1976)
Phleum pratense (timothy)	Mature seed and root	Calli	Atkin and Barton (1973)
Phragmites communis	Stem segment	Plantlets	Sangwan and Gorenflot (1975)
Poa pratensis (Kentucky bluegrass)	Mature embryo	Plants	McDonnell and Conger (1984)
	Mature embryo	Plants	Boyd and Dale (1986)
	Inflorescence and mature seed	Plants	Van der Valk et al. (1989)
	Mature seed	Plantlets	Van Ark et al. (1991)
	Mature seed	Plants	Van der Valk et al. (1995)

Poa trivialis (roughstalk bluegrass)	Mature seed and root	Calli	Atkin and Barton (1973)
Puccinellia limosa (salt marsh grass)	Mature seed	Plantlets	Heszky et al. (1989)
Schizachyrium scoparium (little bluestem)	Immature inflorescence	Plants	Songstad et al. (1986)
Setaria italica (foxtail millet)	Shoot apex	Fertile plants	Osuna-Avila et al. (1995)
Sorghastrum nutans (Indiangrass)	Immature inflorescence	Plantlets	Chen et al. (1979)
Stenotaphrum secundatum (St. Augustine grass)	Immature embryo	Plants	Kuo and Smith (1993)
Stipa viridula (green needlegrass)	Immature inflorescence	Plants	Lo et al. (1980)
Vetiveria zizanioides (vetiver)	Leaf base	Plants	Mucciarelli et al. (1993)
Zoysia japonica (Japanese lawngrass)	Mature seed	Plants	Asano (1989)

vitro and required subculture at 20- to 25-day intervals. The cultures were maintained for over 2 years, retaining characteristics indicative of their endosperm origin, and did not show any morphogenic response.

First callus cultures were initiated and established in *F. arundinacea*, *F. pratensis*, *F. rubra* and *L. multiflorum* from caryopses, isolated mature embryos, whole seedlings, primary root and leaf-tip explants from seedlings (Atkin and Barton 1973; Conger et al. 1978, 1982). Modified LS (Linsmaier and Skoog 1965) and MS (Murashige and Skoog 1962) basal media supplemented with different auxins, i.e. 2,4-D, 2,4,5,-trichlorophenoxyacetic acid (2,4,5,-T), α-2,4,5-trichlorophenoxypropionic acid (silvex), p-chlorophenoxyacetic acid (pCPA), 3,6-dichloro-*o*-anisic acid (dicamba), 4-amino-3,5,6,-trichloropicolinic acid (picloram), α-naphthaleneacetic acid (NAA) and indole-3-acetic acid (IAA), were compared in these studies. Differential responses were found for maximum callus induction and growth with regard to the different auxins, auxin concentrations and explant types. Dicamba and picloram were more effective than 2,4-D in inducing callus in tall fescue. Production of callus from leaf tips was highly variable, and the frequency of callus induction decreased from 86 to 0% as the leaf length increased from 9 to 29 mm in Italian ryegrass. However, callus induction was 88% or above if the node was included in the explant regardless of the initial leaf length. Generally, higher auxin concentrations were required for maximum callus growth of leaf-tip explants than for that of embryo explants (Conger et al. 1978, 1982). Limited organogenesis was observed in most of these cultures and no regeneration of whole plants was achieved. Root primordia that grew into visible roots were found, and shoot primordia, which fail to differentiate into shoots, were described only in the callus cultures of *L. multiflorum* (Atkin and Barton 1973).

In most cases, for the establishment of regenerable callus cultures in fescues and ryegrasses (Table 3.1), calli are induced from meristematic tissues by plating explants from young leaves, apical meristems, immature inflorescences, and immature or mature embryos (Fig. 3.1A) on MS or LS media supplemented with 2,4-D. The continued proliferation and growth of the induced calli in subsequent subcultures is generally achieved by reducing the levels of 2,4-D in the proliferation medium (Fig. 3.1B). Plant regeneration from callus cultures through organogenesis and somatic embryogenesis is promoted by further reducing (or omitting) the levels of 2,4-D, by reducing the strength of the culture medium (e.g. half-strength), and by adding cytokinins such as kinetin or zeatin (Fig. 3.1C, D).

In tall fescue, *F. arundinacea*, whole plants have been regenerated from callus cultures induced from isolated mature embryos and caryopses (Lowe and Conger 1979; Reed and Conger 1985; Eizenga 1989; Takamizo et al. 1994), isolated immature embryos (Takamizo et al. 1994), leaf bases (Kearney et al. 1991), immature inflorescences (Eizenga and Dahleen 1990), and apical meristems (Garcia et al. 1994; Table 3.1). In these studies, MS and SH (Schenk and Hildebrandt 1972) basal media supplemented with 2,4-D in the concentration

Fig. 3.1A–J. Plant regeneration from embryogenic calli of tall fescue (*Festuca arundinacea*) and apomictic dallisgrass (*Paspalum dilatatum*). **A** Callus formation from tall fescue seeds 2 months after plating onto proliferation medium. **B** Detailed view of embryogenic callus from **A**. **C** In vitro regeneration of green plantlets from embryogenic calli of tall fescue. **D** Tall fescue plants derived from callus cultures, growing in the field. **E** Calli induced from seeds of apomictic dallisgrass 3 months after plating seeds. **F** Detailed view of embryogenic callus from **E**. **G** Differentiation of preformed somatic embryos on embryogenic seed-derived callus of apomictic dallisgrass. **H** Rooted in vitro-grown plantlets regenerated from callus cultures of apomictic dallisgrass. **I** Plants from dallisgrass regenerated from callus, growing in the greenhouse. **J** Inflorescence of callus-derived dallisgrass plant

range of 2–10 mg/l were used for callus induction. Depending on the explant and cultivar used, frequencies of callus induction varied between 2 and over 95% (Lowe and Conger 1979; Eizenga and Dahleen 1990; Takamizo et al. 1994). Differentiation of roots, shoots and somatic embryos was observed in the tall fescue callus cultures. Lowe and Conger (1979) reported 45% of the mature embryo-derived calli producing roots, and 18% producing shoots. The frequency of mature embryo-derived calli forming shoots was found to decrease with increasing numbers of subcultures (Lowe and Conger 1979). In vitro-grown plantlets were easily established from the calli showing shoot formation. Albino plantlets were recovered in 12% of the cases. Regenerated green tall fescue plantlets were successfully established in soil (Lowe and Conger 1979). Significant differences in plant regeneration from callus cultures were reported among different tall fescue cultivars (Takamizo et al. 1994). This comparative study of 17 different tall fescue cultivars revealed frequencies of plant regeneration from mature and immature embryo-derived calli varying between 0 and 78%. The mean frequency of plant regeneration from immature embryo-derived calli (77.5%) was higher than that from mature seed-derived calli (8.4%). Tall fescue cultivars bred from Mediterranean ecotypes showed a lower regeneration frequency irrespective of the explants used for callus induction (Takamizo et al. 1994).

Callus cultures of tall fescue have also proven valuable tools in studies relating to the tall fescue endophytic fungus *Neotyphodium* (formerly *Acremonium*) *coenophialum*. Conger and McDaniel (1983) have used 28-day-old callus cultures to screen tall fescue seed samples for the presence of the endophyte. This technique provided an alternative to the previously used histological procedures or enzyme-linked immunosorbent assays, which are more laborious or dependent on the examination of actively growing plants (Conger and McDaniel 1983). Furthermore, protocols for the infection of tall fescue with *N. coenophialum* using callus cultures have been developed (Johnson et al. 1986; Kearney et al. 1991). Callus cultures derived from endophyte-free tall fescue have been inoculated with mycelium of the endophytic fungus. Up to 17% of the plants regenerated from the inoculated embryogenic callus cultures have proven to be infected with the endophyte. Assuming a genetic uniformity of the tall fescue plants recovered from single genotype-derived callus cultures, these procedures allow the infection of "clonal" plants with different isolates of the endophyte to study the individual contributions of the endophyte and the plant genotype towards traits associated with endophyte-infected tall fescue plants (Kearney et al. 1991). This technique provided an alternative to the previously developed procedures which were dependent on the infection of zygotic embryos or seedlings, thus showing genotype variation of the plant material.

In meadow fescue, *F. pratensis*, non-regenerable callus cultures of moderate growth rate were obtained from root explants plated on LS medium supplemented with 5 mg/l 2,4-D (Atkin and Barton 1973). Regenerable callus cultures from meadow fescue were reported by Wang et al. (1993a) for the first time. Calli were induced from mature seeds plated on modified MS medium supplemented with 5 mg/l 2,4-D. Frequencies of plant regeneration from the mature embryo-derived calli of *F. pratensis* varied between 3 and 20% (Wang et al. 1993a).

The initiation and establishment of callus cultures and subsequent plant regeneration have also been achieved in red fescue, *F. rubra* (Atkin and Barton 1973; Torello et al. 1984, 1985; Zaghmout and Torello 1988, 1992a; Spangenberg et al. 1994). Non-regenerable calli of moderate growth rate were induced from pericycle cells of root explants from red fescue seedlings (Atkin and Barton 1973). Regenerable callus cultures differentiating shoots and roots were established for two cultivars of red fescue using mature seeds as initial explants for callus induction (Torello et al. 1984, 1985). The established callus cultures retained their regenerable character over a period of 18 months in culture. Frequencies of embryogenic callus were between 18 and 26%. A total of 122 plants were regenerated from 20 embryogenic calli of one cultivar, and some in vitro regenerated plantlets survived their transfer to soil. Albino plantlets were found in 8 to 18% of the cases (Torello et al. 1984, 1985). Regenerable long-term callus cultures of red fescue have been maintained for over 5 years (Zaghmout and Torello 1992a). Additions of activated charcoal (Zaghmout and Torello 1988) and of elevated sucrose levels (Zaghmout and Torello 1992a) to the maintenance media before regeneration have been shown to result in improved morphogenic capacities of these long-term red fescue in vitro cultures.

In Italian ryegrass, *L. multiflorum*, and interspecific *Lolium* hybrids and *Lolium-Festuca* hybrids involving Italian ryegrass, callus cultures have been established from a wide range of explants including mature seeds (Atkin and Barton 1973; Conger et al. 1978; Wang et al. 1993b), isolated mature and immature embryos (Conger et al. 1978, 1982; Dale 1980), root explants (Atkin and Barton 1973; Jackson et al. 1986; Jackson and Dale 1988), internode and peduncle explants (Kasperbauer et al. 1979), immature inflorescences and nodes (Dale et al. 1981; Creemers-Molenaar et al. 1988), meristem tips (Jackson and Dale 1989), and leaf bases and leaf tips (Conger et al. 1982; Joarder et al. 1986; Jackson and Dale 1988, 1989). In most of these studies, modified basal MS media supplemented with 2,4-D were used for callus induction.

Plants were regenerated first from callus cultures initiated from cultured immature embryos of *L. multiflorum* (Dale 1980), from immature embryos of interspecific *Lolium* hybrids involving Italian ryegrass (Ahloowalia 1975, 1983), and from peduncle and internode explants of *L. multiflorum* × *F. arundinacea* F_1 hybrids (Kasperbauer et al. 1979). Depending on the plant genotype, frequencies of embryogenic callus formation from immature zygotic embryos and mature seeds of Italian ryegrass and its interspecific hybrid varied between 0 and 38% (Dale 1980; Wang et al. 1993b). Up to 17% of the preformed somatic embryos differentiated in vitro-grown plantlets, and up to 50% of them were albino (Ahloowalia 1975, 1983; Dale 1980; Wang et al. 1993b).

Different pathways of morphogenesis were reported for callus cultures initiated from immature inflorescences and nodes of greenhouse- and field-grown Italian ryegrass plants (Dale et al. 1981; Creemers-Molenaar et al. 1988). The regeneration pathway was dependent on the explant type, its age, the culture medium used, and the cultivar. Somatic embryogenesis was observed for calli obtained from both explant sources. The frequency of embryogenic callus formation varied between 4 and 17% for the immature inflorescences depending on their developmental stage (Dale et al. 1981). The percentage of explants forming embryogenic calli differed strongly with the cultivar and the growth condition of the donor plants. Frequencies of embryogenic callus formation were higher for inflorescence explants from greenhouse-grown donor plants (Creemers-Molenaar et al. 1988). The embryogenic calli differentiated both green and albino plantlets. Green plantlets were potted in soil and grown to phenotypically normal plants (Dale et al. 1981; Creemers-Molenaar et al. 1988). Up to 50% of the regenerated plantlets were albino, and explants from field-grown plants showed a higher tendency to form albino shoots than explants from greenhouse-grown donor plants (Dale et al. 1981; Creemers-Molenaar et al. 1988).

Plant regeneration from callus cultures initiated from root explants and root tips has been reported for different Italian ryegrass genotypes (Jackson et al. 1986). The frequency of embryogenic callus formation was dependent on the genotype used. Plant genotypes that have proven responsive in the production of embryogenic callus cultures initiated from apical meristems showed similar good performance when root explants were used (Jackson et al. 1986). Plantlets regenerated from 6- to 27-week-old root-derived callus cultures were transferred to soil and grown until maturity under field conditions (Jackson et al. 1986).

The in vitro response of 1-mm explants from leaves of Italian ryegrass has been compared with respect to the leaf segment position, leaf age and in vivo mitotic activity (Joarder et al. 1986). The frequency of callusing explants decreased with distance from the base of the leaf and with leaf age. A high frequency of callus formation was associated with a high mitotic index in the explant, and a decline in mitotic activity was observed in leaves as they aged (Joarder et al. 1986).

In annual ryegrass, *L. rigidum*, organogenic callus cultures from mature seeds and immature embryos were obtained on MS medium supplemented with 2,4-D and casein hydrolysate or yeast extract (Skene and Barlass 1983). Up to 20% of the callus cultures established differentiated shoots and roots, and plants that were regenerated via organogenesis but not via somatic embryogenesis were recovered and established in soil under greenhouse conditions (Skene and Barlass 1983).

In perennial ryegrass, *L. perenne*, following the pioneering work by Norstog (1956) with the establishment of endosperm callus cultures, plant regeneration systems have been developed based on callus cultures initiated from mature seeds (Torello and Symington 1984; Zaghmout and Torello

1992b; Wang et al. 1993b), immature inflorescences (Creemers-Molenaar et al. 1988), mature embryos (Posselt and Altpeter 1994), stem segments (Zaghmout and Torello 1992b), and meristem tips (Olesen et al. 1995). The effects of different levels of 2,4-D and casein hydrolysate on the production of morphogenic callus, callus maintenance and subsequent plant regeneration have been determined for different perennial ryegrass varieties (Torello and Symington 1984; Creemers-Molenaar et al. 1988). Modified MS (Murashige and Skoog 1962) and B5 (Gamborg et al. 1976) basal media were used in these studies. For the induction of embryogenic callus from immature inflorescences of perennial ryegrass the optimal 2,4-D concentration in the culture medium could not be determined. Optimum shoot formation occurred on B5 media supplemented with 0.1 mg/l 6-benzylamino purine (BAP), while root formation was enhanced by the addition of 2,4-D in the concentration range of 0.1–0.5 mg/l. Callus cultures maintained on media containing 10 mg/l 2,4-D were found to have less regeneration potential than cultures maintained with 5 mg/l 2,4-D (Torello and Symington 1984). Plant regeneration was achieved for all cultivars tested. Both green and albino plantlets were recovered, and the frequency of albino plant regeneration depended on the cultivar and increased with increasing age of the callus cultures. Furthermore, when explanted immature inflorescences from greenhouse- and field-grown donor plants were compared, significantly more albino plantlets were regenerated from callus cultures initiated with explants from field-grown plants (Creemers-Molenaar et al. 1988). In this study, higher concentrations of 2,4-D (up to 15 mg/l) resulted in a higher regeneration frequency of albino shoots.

Zaghmout and Torello (1992b) have studied the effect of different media supplements (cytokinins, fluridone, cefotaxime, activated charcoal, silver nitrate) on plant regeneration from 2 month old callus cultures of perennial ryegrass. The number of regenerated plants was increased by the addition of different cytokinins (BAP, kinetin, zeatin) and 0.2% activated charcoal to the media, but all regenerants were albino. The addition of cefotaxime to the callus proliferation medium prior to the transfer of the calli onto medium for regeneration led to an increase in the frequency of green plants regenerated (Zaghmout and Torello 1992b).

Significant genotype effects on frequency of callus induction, callus growth rate, and regeneration capacity in perennial ryegrass callus cultures have been demonstrated (Wang et al. 1993b; Olesen et al. 1995). A thorough evaluation of 21 genotypes of *L. perenne* revealed that differences between genotypes accounted for 40% of the total variation in callus induction and initial callus growth, for 59% of the variation in regeneration frequency, and for 83% of the variation in frequency of green plants regenerated from callus cultures (Olesen et al. 1995). Furthermore, this study indicated significant genotypic correlations between the regeneration capacity in callus cultures, anther cultures (Chap. 4) and suspension cultures (Chap. 5) of perennial ryegrass.

3.3 Somaclonal Variation in *Festuca* and *Lolium*

Somatic cell culture such as callus cultures were initially conceived as tools for clonally propagating outstanding grass plants for breeding programs (Kaul 1990). The recovery of plants from tissue cultures, particularly from callus cultures, that were phenotypically different from the donor plant was thus first regarded as an impediment to clonal propagation and later as a potential source of useful variation.

Evidence of somaclonal variation in cell cultures in *Festuca* and *Lolium* dates back to the first reports on plant regeneration from callus cultures (Ahloowalia 1975). Somaclonal variants have been found among 4000 plants which were regenerated over a 5-year period from a single callus initiated from a triploid (2n = 21) embryo of an interspecific *Lolium* hybrid, *L. multiflorum* (2x) × *L. perenne* (4x) (Ahloowalia 1983). Chromosome counts made on the regenerated plants varied in the range 2n of 15 to 40 chromosomes. The meiotic study performed on some of the regenerated plants revealed the presence of reciprocal chromosomal translocations, inversions, deletions, chromosomal fragmentation and elimination, and polyploidy. Some of the regenerated plants were mosaics with different chromosome counts in one plant. Furthermore, an albino mutant sector of the triploid callus regenerated albino plantlets, whereas the remaining callus led to the recovery of green plants (Ahloowalia 1983). In addition, the green ryegrass plants recovered showed wide variation in leaf size and shape, growth vigour, survival, perenniality and floral development (Ahloowalia 1983). Some of the variants represented combinations of parental traits and novel types of variation, which, according to the author, had not been observed in hybrid ryegrasses that had not been subjected to tissue culture cycles. Seed progeny from the variants also showed the variation (Ahloowalia 1983).

Phenotypic variants have been reported among over 1200 regenerants from callus cultures of *L. multiflorum* × *F. arundinacea* F_1 (2n = 28) hybrids (Kasperbauer et al. 1979). While plants regenerated from first callus subcultures were identical in appearance and chromosome number to the donor plants, increasing the number of subcultures before plant regeneration increased the frequency of variants in leaf width and coarseness and of amphiploid plants with doubled chromosome number (2n = 56) among the regenerants (Kasperbauer et al. 1979).

Somaclonal variation has been detected among regenerants from callus cultures in *F. arundinacea* (Reed and Conger 1985; Eizenga 1989; Dahleen and Eizenga 1990; Eizenga and Dahleen 1990; Eizenga and Cornelius 1991; Garcia et al. 1994; Roylance et al. 1994).

Reed and Conger (1985) reported results from comparative meiotic analyses of tall fescue plants regenerated from seeds and callus cultures, particularly from those cultures in which the mature embryos used for callus induction had been treated with ethylmethane sulfonate (EMS) as chemical mutagen. While no chromosomal abonormalities were found in pollen mother cells of the seed-derived control plants, between 40 and 50% of the tall fescue callus culture regenerants exhibited chromosomal abnormalities such as lagging chromosomes, formation of micronuclei and anaphase bridges, as well as aneuploidy. The tall fescue plants which were regenerated from callus cultures initiated from EMS-treated or untreated embryos did not differ in either frequency or the types of meiotic aberrations observed. Pollen viability was relatively high (>95%) for all three categories of regenerants and thus appeared not to be related to the frequency of aberrations found in the individual regenerated tall fescue plants (Reed and Conger 1985).

Meiotic analyses of 166 tall fescue somaclones regenerated via organogenesis from mature embryo-derived 14-week-old callus cultures and compared to 15 parental clones have been described by Eizenga (1989). The analysis of the parental clones revealed mainly ring bivalent pairing and pollen viability between 76 and 98%. In contrast, the analogous meiotic evaluation of the callus culture-derived somaclones showed 60% euploids, 35% aneuploids with 2n = 38, 39, 40 and 41, and 5% of higher ploidy with 2n = 56, 70, 78 and 80 chromosomes. Abnormal meiotic behaviour, translocations, deletions, increased multivalent frequencies, achiasmatic meiosis and drastically reduced pollen stainability below 10% were observed in some of the somaclones analyzed (Eizenga 1989). Isozyme changes for acid phosphatase-1 (ACP-1), alcohol dehydrogenase 1 (ADH-1), glutamate oxaloacetate transaminase 2 (GOT-2) and 3 (GOT-3), malate dehydrogenase (MDH), 6-phosphogluconate dehydrogenase-1 (6-PGD-1) and phosphoglucoisomerase 2 (PGI-2) were also implicated in these regenerants when compared to the parental clones (Dahleen and Eizenga 1990; Eizenga and Cornelius 1991). The culture-induced isozyme variation, however, appeared to alter the distribution of the isozyme phenotypes, but was not directly correlated with chromosome abnormalities (Eizenga and Cornelius 1991). The presence of reproductive disturbances with abnormal microsporogenesis affecting pollen sterility, chromosomal aberrations such as paracentric inversions, sticky chromosomes, reciprocal translocations and deletions, as well as instabilities at isozyme loci have been confirmed in tall fescue plants regenerated from callus cultures initiated from shoot tips, even after only 1 month in callus culture (Garcia et al. 1994).

In contrast to the frequent aberrations observed among tall fescue somaclones from callus cultures initiated from mature embryos and shoot tips, Eizenga and Dahleen (1990) reported the lack of significant cytological changes in plants regenerated probably via somatic embryogenesis from 4-week-old cultures initiated from immature inflorescences of tall fescue. From a total of 210 tissue culture regenerants, 6% were albino. The cytological evaluation of 95 green tall fescue regenerants revealed that all plants had 21 bivalents and pollen viability ranging from 43 to 98%. Only differences in the frequencies of ring to rod bivalents indicating differences in chromosome pairing were detected among the 95 regenerants analyzed. Zymograms for seven enzyme systems performed for all 95 tall fescue plants showed no differences from that of the parental genotype, thus confirming the somatic origin of the tissue culture regenerated plants (Eizenga and Dahleen 1990).

Roylance et al. (1994) have found quantitative traits affected by somaclonal variation in seven tall fescue plants regenerated from callus cultures via somatic embryogenesis using a protocol designed to reduce somaclonal variation (Kearney et al. 1991). The tall fescue tissue culture regenerants were screened under field conditions for somaclonal variation in phenological development, total biomass, vegetative- and flag-leaf width/length ratios, pollen viability and seed yield for 2 years. No differences were found between the tissue culture regenerants and control tall fescue plants with respect to dates of panicle emergence, flowering, seed maturity and pollen viability. However, significant differences were detected for total biomass, flag-leaf width/length ratio, vegetative-leaf width/length ratio and seed yield (Roylance et al. 1994).

Similarly, contrasting reports describing either the lack of obvious morphological abnormalities and chromosome changes or the presence of cytological and morphological variation among plants regenerated from callus cultures have been published for *L. multiflorum* (Dale 1980; Jackson et al. 1986; Jackson and Dale 1988, 1989).

Dale (1980) described the regeneration of green and albino plantlets from callus cultures initiated from immature embryos of Italian ryegrass, with 50% of the regenerants albino. However, the evaluation of 15 green plantlets grown until maturity revealed no somaclonal variation and all regenerants had the expected normal chromosome count 2n = 14 (Dale 1980).

Evidence for the occurrence of somaclonal variants was obtained from the analysis of plants regenerated from callus cultures initiated from roots, shoot apices and leaf explants for several genotypes of *L. multiflorum* and *L. temulentum* (Jackson et al. 1986; Jackson and Dale 1988, 1989). Frequencies of albino regenerants varied between 2 and 18% depending on the genotype, and the increase in culture temperature was found to increase the frequency of albino plant regeneration from callus cultures of one Italian ryegrass genotype (Jackson and Dale 1988). Cytological variation was also observed in this study, and while all regenerants from callus cultures initiated from a diploid genotype had the normal 2n = 14 chromosome count, plants regenerated from callus derived from tetraploid cultures had lost up to three chromosomes. Chromosome counts performed for 53 Italian ryegrass plants that were regenerated from 6- to 27-week-old callus cultures initiated from root explants revealed that one plant was a tetraploid, while all the others had the normal 2n = 14 count. These plants displayed significant differences in heading date, plant height and width at heading in field trials assessed at an early stage, so some of the variation observed might have been due to physiological factors (Jackson et al. 1986).

Jackson and Dale (1989) convincingly documented variation in plant height, flowering time, chromosome number and isozyme variation at the superoxide dismutase (SOD) locus among 65 regenerants from shoot apices-derived callus cultures of Italian ryegrass. In the same study, regenerants from immature embryo-derived callus cultures of the inbreeding species *L. temulentum* were considered. Progeny of the regenerated *L. temulentum* plants showed variation in height, ear length, flowering date and chlorophyll content (Jackson and Dale 1989).

Although variation among callus culture-derived plants has been clearly demonstrated in *Festuca* and *Lolium* species, there have been so far very limited attempts to make use of the advocated application of somaclonal variation associated with in vitro selection for desirable mutations at the cellular level in order to improve fescues and ryegrasses. In vitro selection of callus cultures from two populations of *L. perenne* and the toxin deoxynivalenol (DON) produced by *Michrodochium nivale* as selective agent have been tested as a means to regenerate plants with snow mould resistance (Posselt and Altpeter 1994). Perennial ryegrass regenerants from in vitro selection were artificially inoculated with *M. nivale,* and significant improvement in snow mould tolerance in one high DON variant could be found. A high level of DON resistance was observed in sexual progenies from the selected variants, indicative of heritable factors involved in toxin resistance. However, no close correlation with pathogen resistance was attained (Posselt and Altpeter 1994).

3.4 Regeneration from Callus Cultures in Other Grasses

Callus cultures have been established in over 60 species of forage and turf grasses belonging to over 25 genera outside the *Festuca-Lolium* complex (Table 3.1). Plants have been regenerated from callus cultures in different grass species belonging to the genera *Agropyron* (Lo et al. 1980; Gyulai et al. 1992), *Agrostis* (Atkin and Barton 1973; Krans et al. 1982), *Alopecurus* (Lo et al. 1980), *Andropogon* (Chen et al. 1977), *Bothriochloa* (Metzinger et al. 1987; Franklin et al. 1990), *Bromus* (Lo et al. 1980), *Cenchrus* (Ross et al. 1995), *Cynodon* (Ahn et al. 1985), *Dactylis* (Conger et al. 1982), *Digitaria* (Gonzales and Franks 1987), *Echinochloa* (Takahashi et al. 1984; Wang and Yan 1984; Samantaray et al. 1995, 1997), *Eleusine* (Rangan 1976), *Elymus* (Park and Walton 1989a,b), *Eragrostis* (Echenique et al. 1996), *Panicum* (Lu and Vasil 1981; Rangan and Vasil 1983; Denchev and Conger 1994), *Paspalum* (Bovo and Mroginski 1986; Akashi and Adachi 1992), *Pennisetum* (Haydu and Vasil 1981; Vasil and Vasil 1981), *Poa* (McDonnell and Conger 1984), *Setaria* (Osuna-Avila et al. 1995), *Stenotaphrum* (Kuo and Smith 1993), *Vetiveria* (Mucciarelli et al. 1993) and *Zoysia* (Asano 1989).

For the initiation of these regenerable callus cultures in a wide range of forage and turf grasses, different explants containing meristematic cells or showing a developmental proximity to the embryo have been used, e.g.

isolated mature and immature embryos (Vasil and Vasil 1981; Lu and Vasil 1982; Boyd and Dale 1986), caryopses (Conger and Carabia 1978; Eapen and George 1990; Franklin et al. 1990), immature inflorescences (Chen et al. 1977; Lo et al. 1980; Vasil and Vasil 1981; Wang and Vasil 1982; Tyagi et al. 1985; Bovo and Mroginski 1986), shoot apices (Wu and Antonovics 1978; Osuna-Avila et al. 1995), young leaf bases (Haydu and Vasil 1981; Lu and Vasil 1981; Conger et al. 1983; Hanna et al. 1984; Mohanty et al. 1985; Mucciarelli et al. 1993), and mesocotyls and leaf stem segments (Rangan 1976; Cobb et al. 1985; Fladung and Hesselbach 1986; Shatters et al. 1994).

Established callus cultures of forage and turf grasses have been shown to produce shoots, roots and somatic embryos that finally lead to the recovery of whole plants (Ahloowalia 1984; Table 3.1). Regeneration from somatic cells in grass callus cultures was found to take place via organogenesis by adventitious shoot formation and subsequent root formation from meristematic regions or by somatic embryogenesis through the formation of embryoids that were capable of germination. Furthermore, callus initiation and growth in vitro and subsequent somaclonal variation occurred, with plants regenerated via somatic embryogenesis usually being less variable than plants regenerated via shoot and root morphogenesis.

Taken together, the above studies clearly illustrate the fact that calli produced from different forage and turf grasses and different tissue explants may show considerable variation in morphology, developmental capabilities and genetic stability. Somaclonal variation can be observed among regenerants from callus cultures for various morphological and agronomic traits of forage and turf grasses (Duncan 1996). However, the potential to exploit the genetically heritable variation to develop novel breeding material in forage and turf grasses has so far led to the release of only a single cultivar, namely *Cynodon dactylon* cv. Brazos R3 with improved fall armyworm resistance (Croughan et al. 1994). Nevertheless, callus cultures show specific potential to generate useful variation, particularly in apomictic grass species such as apomictic dallisgrass, where protocols for plant regeneration have been established (Fig. 3.1E–J) and the reproductive mode represents a barrier to recombination yet offers unique opportunities to fix agronomically interesting variant genotypes arising from tissue cultures.

3.5 Summary and Conclusions

Efficient and reproducible protocols for the induction of callus from young meristematic tissues and the establishment of regenerable callus cultures until regeneration of mature soil-grown plants are in place for different *Festuca* and *Lolium* species.

Generally, isolated immature or mature embryos or intact seeds are placed on solidified simple basal media supplemented with a phenoxy herbicide with auxin-like activity to induce callus formation. Different callus types are produced, and embryogenic calli are selected and proliferated to establish regenerable callus cultures. This selection and preferential maintenance of embryogenic calli is critical to the retention of long-term morphogenic potential of the established callus cultures. Differentiation of preformed somatic embryos occurs when embryogenic calli are transferred to regeneration media. Since a high auxin concentration inhibits embryo germination, levels of auxin are lowered or omitted for plant regeneration.

The use of embryos as explants for callus induction, establishment of callus cultures and subsequent plant regeneration is technically preferred. However, due to their zygotic origin, the resulting cultures represent sexual progenies and not clones of performance-tested individuals. The use of explants such as immature inflorescences, leaf bases or roots from mature, outstanding tested plant material is recommended for the establishment of regenerable callus cultures to clonally propagate selected genotypes. A regeneration system through somatic embryogenesis and a short tissue culture phase is recommended to reduce effects of somaclonal variation when clonal propagation is aimed.

Somaclonal variation, i.e. the recovery of genetic changes in plants regenerated from tissue culture, has been observed in regenerants from callus cultures in *Festuca* and *Lolium* species. Few authors, however, have attempted to study the inheritance of the variation observed. Although plant regeneration from callus cultures offers an opportunity to uncover natural variability and to potentially use this variability as a random, undirected form of mutation breeding for the development of improved cultivars in fescues and ryegrasses, no useful variant has yet been released.

References

Ahloowalia BS (1975) Regeneration of ryegrass plants in tissue culture. Crop Sci 15: 449–452

Ahloowalia BS (1983) Spectrum of variation in somaclones of triploid ryegrass. Crop Sci 23: 1141–1147

Ahloowalia BS (1984) Forage grasses. In: Ammirato PV, Evans DA, Sharp WR, Yamada Y (eds) Handbook of plant cell culture, vol 3. Macmillan, New York, pp 91–125

Ahn BJ, Huang FH, King JW (1985) Plant regeneration through somatic embryogenesis in common bermudagrass tissue culture. Crop Sci 25: 1107–1109

Ahn BJ, Huang FH, King JW (1987) Regeneration of bermudagrass cultivars and evidence of somatic embryogenesis. Crop Sci 27: 594–597

Akashi R, Adachi T (1991) High frequency somatic formation in cultures of immature embryos of guineagrass, *Panicum maximum* Jacq. Jpn J Breed 41: 85–93

Akashi R, Adachi T (1992) Somatic embryogenesis and plant regeneration from cultured immature inflorescences of apomictic dallisgrass (*Paspalum dilatatum* Poir.). Plant Sci 82: 213–218

Akashi R, Hashimoto A, Adachi T (1993) Plant regeneration from seed-derived embryogenic callus and cell suspension cultures of bahiagrass (*Paspalum notatum*). Plant Sci 90: 73–80

Artunduaga IR, Taliaferro CM, Johnson BL (1988) Effects of auxin concentration on induction and growth of embryogenic callus from young inflorescence explants of Old World bluestem (*Bothriochloa spp.*) and bermuda (*Cynodon* spp.) grasses. Plant Cell Tissue Organ Cult 12: 13–19
Artunduaga IR, Taliaferro CM, Johnson BB (1989) Induction and growth of callus from immature inflorescences of "Zebra" bermudagrass as affected by casein hydrolysate and 2,4-D concentration. In vitro Cell Dev Biol 25: 753–756
Asano Y (1989) Somatic embryogenesis and protoplast culture in Japanese lawngrass (*Zoysia japonica*). Plant Cell Rep 8: 141–143
Atkin RK, Barton GE (1973) The establishment of tissue cultures of temperate grasses. J Exp Bot 24: 689–699
Blanche FC, Krans JV, Coats GE (1986) Improvement in callus growth and plantlet formation in creeping bentgrass. Crop Sci 26: 1245–1248
Bovo OA, Mroginski LA (1986) Tissue culture in *Paspalum* (Gramineae): plant regeneration from cultured inflorescences. J Plant Physiol 124: 481–492
Boyd LA, Dale PJ (1986) Callus production and plant regeneration from mature embryos of *Poa pratensis* L. Plant Breed 97: 246–254
Chandler SF, Vasil IK (1984) Optimization of plant regeneration from long term embryogenic callus cultures of *Pennisetum purpureum* Schum. (Napier grass). J Plant Physiol 117: 147–156
Chen CH, Stenberg NE, Ross JG (1977) Clonal propagation of big bluestem by tissue culture. Crop Sci 17: 847–850
Chen CH, Lo PF, Ross JG (1979) Regeneration of plantlets from callus cultures of indiangrass. Crop Sci 19: 117–118
Cobb BG, Vanderzee D, Loescher WH, Kennedy RA (1985) Evidence for plantlet regeneration via somatic embryogenesis in the grasses *Echinochloa muricata* and *E. crus-galli* var. *oryzicola*. Plant Sci 40: 121–127
Conger BV, Carabia JV (1978) Callus induction and plantlet regeneration in orchardgrass. Crop Sci 18: 157–159
Conger BV, McDaniel JK (1983) Use of callus cultures to screen tall fescue seed samples for *Acremonium coenophialum*. Crop Sci 23: 172–174
Conger BV, Carabia JV, Lowe KW (1978) Comparison of 2,4-D and 2,4,5-T on callus induction and growth in three Gramineae species. Environ Exp Bot 18: 163–168
Conger BV, Hilenski LL, Lowe KW, Carabia JV (1982) Influence of different auxins at varying concentrations on callus induction and growth from embryo and leaf-tip explants in Gramineae. Environ Exp Bot 22: 39–48
Conger BV, Hanning GE, Gray DJ, McDaniel JK (1983) Direct embryogenesis from mesophyll cells of orchardgrass. Science 221: 850–851
Creemers-Molenaar J, Loeffen JPM, Van der Valk P (1988) The effect of 2,4-dichlorophenoxyacetic acid and donor plant environment on plant regeneration from immature inflorescence-derived callus of *Lolium perenne* L. and *Lolium multiflorum* L. Plant Sci 57: 165–172
Croughan SS, Quisenberry SS, Eichhorn MM, Colyer PD, Brown TF (1994) Registration of Brazos-R3 bermudagrass germplasm. Crop Sci 34: 542
Dahleen LS, Eizenga GC (1990) Meiotic and isozymic characterization of plants regenerated from euploid and selfed monosomic tall fescue embryos. Theor Appl Genet 79: 39–44
Dale PJ (1980) Embryoids from cultured immature embryos of *Lolium multiflorum*. Z Pflanzenphysiol 100: 73–77
Dale PJ, Thomas E, Brettell RIS, Wernicke W (1981) Embryogenesis from cultured immature inflorescences and nodes of *Lolium multiflorum*. Plant Cell Tissue Organ Cult 1: 47–55
Denchev PD, Conger BV (1994) Plant regeneration from callus cultures of Switchgrass. Crop Sci 34: 1623–1627
Denchev PD, Conger BV (1995) In vitro culture of switchgrass: Influence of 2,4-D and picloram in combination with benzyladenine on callus initiation and regeneration. Plant Cell Tissue Organ Cult 40: 43–48

Duncan RR (1996) Tissue culture-induced variation and crop improvement. Adv Agron 58: 201–240

Eapen S, George L (1989) High frequency plant regeneration through somatic embryogenesis in finger millet (*Eleusine coracana* Gaertn). Plant Sci 61: 27–130

Eapen S, George L (1990) Influence of phytohormones, carbohydrates, amino acids, growth supplements and antibiotics on somatic embryogenesis and plant differentiation in finger millet. Plant Cell Tissue Organ Cult 22: 87–93

Echenique V, Polci P, Mroginski L (1996) Plant regeneration in weeping lovegrass (*Eragrostis curvula*) through inflorescence culture. Plant Cell Tissue Organ Cult 46: 123–130

Eizenga GC (1989) Meiotic analysis of tall fescue somaclones. Genome 32: 373–379

Eizenga GC, Cornelius PL (1991) Comparison of the isozyme variation in tall fescue parents and their somaclones. Euphytica 51: 249–256

Eizenga GC, Dahleen LS (1990) Callus production, regeneration and evaluation of plants from cultured inflorescences of tall fescue (*Festuca arundinacea* Schreb.). Plant Cell Tissue Organ Cult 22: 7–15

Evans DA (1989) Somaclonal variation – genetic basis and breeding applications. Trends Genet 5: 46–50

Evans DA, Sharp WR, Medina-Filho HP (1984) Somaclonal and gametoclonal variation. Am J Bot 71: 759–774

Fladung M, Hesselbach J (1986) Callus induction and plant regeneration in *Panicum bisulcatum* and *Panicum milioides*. Plant Cell Rep 3: 169–173

Franklin CI, Trieu TN, Gonzales RA (1990) Plant regeneration through somatic embryogenesis in the forage grass Caucasian bluestem (*Bothriochloa caucasica*). Plant Cell Rep 9: 443–446

Gamborg OL, Murashige T, Thorpe TA, Vasil IK (1976) Plant tissue culture media. In Vitro 12: 473–478

Garcia A, Dalton SJ, Humphreys MO (1994) Reproductive disturbances and phosphoglucoisomerase instability in *Festuca arundinacea* (tall fescue) plants regenerated from callus and cell suspension cultures. Heredity 73: 355–362

Gavin AL, Conger BV, Trigiano RN (1989) Sexual transmission of somatic embryogenesis in *Dactylis glomerata*. Plant Breed 103: 251–254

Gonzales RA, Franks E (1987) Induction of morphogenically competent callus and suspension cell cultures from leaf explants and mature seeds of *Digitaria sanguinalis*. In Vitro Cell Dev Biol 23: 581–584

Gray DJ (1990) Somatic cell culture and embryogenesis in the Poaceae. In: Kasperbauer MJ (ed) Biotechnology in tall fescue improvement. CRC Press, Boca Raton, pp 25–57

Gyulai G, Janovszky J, Kiss E, Lelik L, Csillag A, Heszky LE (1992) Callus initiation and plant regeneration from inflorescence primordia of the intergeneric hybrid *Agropyron repens* (L.) Beauv. × *Bromus inermis* Leyss. cv. *nanus* on a modified nutritive medium. Plant Cell Rep 11: 266–269

Hanna WW, Lu C, Vasil IK (1984) Uniformity of plants regenerated from somatic embryos of *Panicum maximum* Jacq. (Guinea grass). Theor Appl Genet 67: 155–159

Hanning GE, Conger BV (1982) Embryoid and plantlet formation from leaf segments of *Dactylis glomerata* L. Theor Appl Genet 63: 155–159

Hanning GE, Conger BV (1986) Factors influencing somatic embryogenesis from cultured leaf segments of *Dactylis glomerata*. J Plant Physiol 123: 23–29

Haydu Z, Vasil IK (1981) Somatic embryogenesis and plant regeneration from leaf tissues and anthers of *Pennisetum purpureum* Schum. Theor Appl Genet 59: 269–273

Heszky LE, Binh DQ, Kiss E, Gyulai G (1989) Increase of green plant regeneration efficiency by callus selection in *Puccinellia limosa* (Schur.) Holmbg. Plant Cell Rep 8: 174–177

Heyser JW, Nabors MW (1982) Regeneration of proso millet from embryogenic calli derived from various plant parts. Crop Sci 22: 1070–1074

Jackson JA, Dale PJ (1988) Callus induction, plant regeneration and an assessment of cytological variation in regenerated plants of *Lolium multiflorum* L. J Plant Physiol 132: 351–355

Jackson JA, Dale PJ (1989) Somaclonal variation in *Lolium multiflorum* L. and *L. temulentum* L. Plant Cell Rep 8: 161–164

Jackson JA, Dalton SJ, Dale PJ (1986) Plant regeneration from root callus in the forage grass *Lolium multiflorum*. In: Withers LA, Alderson PJ (eds) Plant tissue culture and its agricultural applications. Butterworths, London, pp 85–89

Joarder OI, Joarder NH, Dale PJ (1986) In vitro response of leaf tissues from *Lolium multiflorum* – a comparison with leaf segment position, leaf age and in vivo mitotic activity. Theor Appl Genet 73: 286–291

Johnson MC, Bush LP, Siegel MR (1986) Infection of tall fescue with *Acremonium coenophialum* by means of callus culture. Plant Dis 70: 380–382

Kasperbauer MJ, Buckner RC, Bush LP (1979) Tissue culture of annual ryegrass × tall fescue hybrids: callus establishment and plant regeneration. Crop Sci 19: 457–460

Kaul K (1990) Potential biotechnological approaches. In: Kasperbauer MJ (ed) Biotechnology in tall fescue improvement. CRC Press, Boca Raton, pp 13–23

Kearney JF, Parrot WA, Hill NS (1991) Infection of somatic embryos of tall fescue with *Acremonium coenophialum*. Crop Sci 31: 979–984

King PJ, Potrykus I, Thomas E (1978) In vitro genetics of cereals: problems and perspectives. Physiol Veg 16: 381–399

Krans JV, Henning VT, Torres KC (1982) Callus induction, maintenance and plantlet regeneration in creeping bentgrass. Crop Sci 22: 1193–1197

Kuo YJ, Smith MAL (1993) Plant regeneration from St. Augustinegrass immature embryo-derived callus. Crop Sci 33: 1394–1396

Larkin PJ, Scowcroft WR (1981) Somaclonal variation – a novel source of variability from cell cultures for plant improvement. Theor Appl Genet 60: 197–214

Linsmaier EM, Skoog F (1965) Organic growth factor requirement for tobacco tissue cultures. Physiol Plant 18: 100–127

Lo PF, Chen CH, Ross JG (1980) Vegetative propagation of temperate forage grasses through callus culture. Crop Sci 20: 363–367

Lorz H, Gobel E, Brown P (1988) Advances in tissue culture and progress towards genetic transformation of cereals. Plant Breed 100: 1–25

Lowe KW, Conger BV (1979) Root and shoot formation from callus cultures of tall fescue. Crop Sci 19: 397–400

Lu CY, Vasil IK (1981) Somatic embryogenesis and plant regeneration from leaf tissues of *Panicum maximum* Jacq. Theor Appl Genet 59: 275–280

Lu CY, Vasil IK (1982) Somatic embryogenesis and plant regeneration in tissue cultures of *Panicum maximum* Jacq. Am J Bot 69: 77–81

Marousky FJ, West SH (1990) Somatic embryogenesis and plant regeneration from cultured mature caryopses of bahiagrass (*Paspalum notatum* Flugge). Plant Cell Tissue Organ Cult 20: 125–129

McDonnell RE, Conger BV (1984) Callus induction and plantlet formation from mature embryo explants of Kentucky bluegrass. Crop Sci 24: 573–578

Metzinger BD, Taliaferro CM, Johnson BB, Mitchell ED (1987) In vitro regeneration of apomictic bluestem grasses. Plant Cell Tissue Organ Cult 10: 31–38

Mohanty BD, Gupta SD, Ghosh PD (1985) Callus initiation and plant regeneration in ragi (*Eleusine coracana* Gaertn.). Plant Cell Tissue Organ Cult 5: 147–150

Morrish F, Vasil V, Vasil IK (1987) Developmental morphogenesis and genetic manipulation in tissue and cell cultures of the Gramineae. Adv Genet 24: 431–499

Morrish FM, Hanna WW, Vasil IK (1990) The expression and perpetuation of inherent somatic variation in the regenerants from embryogenic cultures of *Pennisetum glaucum* (L.) R. Br. (pearl millet). Theor Appl Genet 80: 409–416

Mucciarelli M, Gallino M, Scannerini S, Maffei M (1993) Callus induction and plant regeneration in *Vetiveria zizanioides*. Plant Cell Tissue Organ Cult 35: 267–271

Murashige T, Skoog F (1962) A revised medium for rapid growth and bioassays with tobacco tissue culture. Physiol Plant 15: 473–497

Nagarathna KC, Shetty SA, Harinarayana G, Shetty HS (1993) Selection for downy mildew resistance from the regenerants of pearl millet. Plant Sci 90: 53–61

Norstog KJ (1956) Growth of rye-grass endosperm in vitro. Bot Gaz 117: 253–259

Olesen A, Storgaard M, Madsen S, Andersen SB (1995) Somatic in vitro culture response of *Lolium perenne* L.: genetic effects and correlations with anther culture. Euphytica 86: 199–209

Osuna-Avila P, Nava-Cedillo A, Jofre-Garfias AE, Cabrera-Ponce JL (1995) Plant regeneration from shoot apex explants of foxtail millet. Plant Cell Tissue Organ Cult 40: 33–35

Park CH, Walton PD (1989a) Embryogenesis and plant regeneration from tissue culture of Canada wildrye, *Elymus canadensis* L. Plant Cell Rep 8: 289–291

Park CH, Walton PD (1989b) Morphology and cytology of tissue culture derived octoploid of *Elymus canadensis* and its selfed progeny. Plant Breed 102: 208–214

Park CH, Kumar PS, Walton PD (1990) Plant regeneration and chromosomal stability in tissue cultures of the hybrids of *Elymus canadensis* with *Psathyrostachys juncea* and *Secale cereale*. Plant Breed 104: 184–189

Pius J, George L, Eapen S, Rao PS (1993) Enhanced plant regeneration in pearl millet (*Pennisetum americanum*) by ethylene inhibitors and cefotaxime. Plant Cell Tissue Organ Cult 32: 91–96

Posselt UK, Altpeter F (1994) Improvement of snow mould resistance by conventional and in vitro techniques. Euphytica 77: 251–255

Rajasekaran K, Schank SC, Vasil IK (1986) Characterization of biomass production, cytology and phenotypes of plants regenerated from embryogenic callus cultures of *Pennisetum americanum* × *P. purpureum* (hybrid triploid napiergrass). Theor Appl Genet 73: 4–10

Rangan TS (1974) Morphogenic investigations on tissue cultures of *Panicum miliaceum*. Z Pflanzenphysiol 72: 456–459

Rangan TS (1976) Growth and plantlet regeneration in tissue cultures of some Indian millets: *Paspalum scrobiculatum* L., *Eleusine coracana* Gaertn. and *Pennisetum typhoideum* Pers. Z Pflanzenphysiol 78: 208–216

Rangan TS, Vasil IK (1983) Somatic embryogenesis and plant regeneration in tissue cultures of *Panicum miliaceum* L. and *Panicum miliare* Lamk. Z Pflanzenphysiol 109: 49–53

Reed JN, Conger BV (1985) Meiotic analysis of tall fescue plants regenerated from callus cultures. Environ Exp Bot 25: 277–284

Ross AH, Manners JM, Birch RG (1995) Embryogenic callus production, plant regeneration and transient gene expression following particle bombardment in the pasture grass, *Cenchrus ciliaris* (Gramineae). Aust J Bot 43: 193–199

Roylance JT, Hill NS, Parrot WA (1994) Detection of somaclonal variation in tissue culture regenerants of tall fescue. Crop Sci 34: 1369–1372

Samantaray S, Rout GR, Das P (1995) In vitro plant regeneration from leaf base and mesocotyl cultures of *Echinochloa colona*. Plant Cell Tissue Organ Cult 40: 37–41

Samantaray S, Rout GR, Das P (1997) Regeneration of plants via somatic embryogenesis from leafbase and leaf tip segments of *Echinochloa colona*. Plant Cell Tissue Organ Cult 47: 119–125

Sangwan RS, Gorenflot R (1975) In vitro culture of *Phragmites* tissues. Callus formation, organ differentiation and cell suspension culture. Z Pflanzenphysiol 75: 256–269

Sankhla A, Davis TD, Sankhla D, Sankhla N, Upadhyaya A, Joshi S (1992) Influence of growth regulators on somatic embryogenesis, plantlet regeneration, and post-transplant survival of *Echinochloa frumentacea*. Plant Cell Rep 11: 368–371

Schenk RU, Hildebrandt AC (1972) Medium and techniques for induction and growth of monocotyledonous and dicotyledonous plant cell cultures. Can J Bot 50: 199–204

Shatters RG, Wheeler RA, West SH (1994) Somatic embryogenesis and plant regeneration from callus cultures of "Tifton 9" Bahiagrass. Crop Sci 34: 1378–1384

Shenoy VB, Vasil IK (1992) Biochemical and molecular analysis of plants derived from embryogenic tissue cultures of napier grass (*Pennisetum purpureum* K. Schum). Theor Appl Genet 83: 947–955

Sivadas P, Kothari SL, Chandra N (1990) High frequency embryoid and plantlet formation from tissue cultures of the finger millet – *Eleusine coracana* (L.) Gaertn. Plant Cell Rep 9: 93–96

Skene KGM, Barlass M (1983) Regeneration of plants from callus cultures of *Lolium rigidum*. Z Pflanzenzuchtg 90: 130–135

Songstad DD, Chen CH, Boe AA (1986) Plant regeneration in callus cultures derived from young inflorescences of little bluestem. Crop Sci 26: 827–829

Spangenberg G, Wang ZY, Nagel J, Potrykus I (1994) Protoplast culture and generation of transgenic plants in red fescue (*Festuca rubra* L.). Plant Sci 97: 83–94

Swedlund B, Vasil IK (1985) Cytogenetic characterization of embryogenic callus and regenerated plants of *Pennisetum americanum* (L.) K. Schum. Theor Appl Genet 69: 575–581

Takahashi A, Sakuragi Y, Kamada H, Ishizuka K (1984) Plant regeneration through somatic embryogenesis in barnyardgrass, *Echinochloa oryzicola* Vasing. Plant Sci Lett 36: 161–163

Takamizo T, Fukase N, Saruwatari Y, Suginobu KI (1994) Genetic variation in plant regeneration from callus culture of tall fescue (*Festuca arundinacea* Schreb.) cultivars. JARQ 28: 200–205

Talwar M, Rashid A (1989) Somatic embryo formation from unemerged inflorescences and immature embryos of a graminaceous crop *Echinochloa*. Ann Bot 64: 195–199

Torello WA, Symington AG (1984) Regeneration from perennial ryegrass callus tissue. HortScience 19: 56–57

Torello WA, Symington AG, Rufner R (1984) Callus initiation, plant regeneration, and evidence of somatic embryogenesis in red fescue. Crop Sci 24: 1037–1040

Torello WA, Rufner R, Symington AG (1985) The ontogeny of somatic embryos from long-term callus cultures of red fescue. HortScience 20: 938–942

Tyagi AK, Bharal S, Rashid A, Maheshwari N (1985) Plant regeneration from tissue cultures initiated from immature inflorescences of a grass, *Echinochloa colonum* (L.) Link. Plant Cell Rep 4: 115–117

Van Ark HF, Zaal MACM, Creemers-Molenaar J, Van der Valk P (1991) Improvement of the tissue culture response of seed-derived callus cultures of *Poa pratensis* L.: effect of gelling agent and abscisic acid. Plant Cell Tissue Organ Cult 27: 275–280

Van der Valk P, Zaal MACM, Creemers-Molenaar J (1989) Somatic embryogenesis and plant regeneration in inflorescence and seed derived callus cultures of *Poa pratensis* L. (Kentucky bluegrass). Plant Cell Rep 7: 644–647

Van der Valk P, Ruis F, Tettelaar-Schrier AM, Van de Velde CM (1995) Optimizing plant regeneration from seed-derived callus cultures of Kentucky bluegrass. The effect of benzyladenine. Plant Cell Tissue Organ Cult 40; 101–103

Vasil IK (1983) Regeneration of plants from single cells of cereals and grasses. In: Lurquin P, Kleinhofs A (eds) Genetic engineering of eukaryotes. Plenum Press, New York, pp 233–252

Vasil IK (1987) Developing cell and tissue culture systems for the improvement of cereals and grass crops. J Plant Physiol 128: 193–218

Vasil V, Vasil IK (1981) Somatic embryogenesis and plant regeneration from tissue cultures of *Pennisetum americanum* and *P. americanum* × *P. purpureum* hybrid. Am J Bot 68: 864–872

Vasil V, Vasil IK (1982) The ontogeny of somatic embryos of *Pennisetum americanum* (L.) K. Schum. I. In cultured immature embryos. Bot Gaz 143: 454–465

Wang DY , Vasil IK (1982) Somatic embryogenesis and plant regeneration from inflorescence segments of *Pennisetum purpureum* Schum. (napier or elephant grass). Plant Sci Lett 25: 147–154

Wang DY, Yan K (1984) Somatic embryogenesis in *Echinochloa crusgalli*. Plant Cell Rep 3: 88–90

Wang ZY, Vallés MP, Montavon P, Potrykus I, Spangenberg G (1993a) Fertile plant regeneration from protoplasts of meadow fescue (*Festuca pratensis* Huds.). Plant Cell Rep 12: 95–100

Wang ZY, Nagel J, Potrykus I, Spangenberg G (1993b) Plants from cell suspension-derived protoplasts in *Lolium* species. Plant Sci 94: 179–193

Wenck AR, Conger BV, Trigiano RN, Sams CE (1988) Inhibition of somatic embryogenesis in orchardgrass by endogenous cytokinins. Plant Physiol 88: 990–992

White PR (1943) A handbook of plant tissue culture. Jaques Cattell, Lancaster.

Wu L, Antonovics J (1978) Zinc and copper tolerance of *Agrostis stolonifera* L. in tissue culture. Am J Bot 65: 268–271

Yamada Y (1977) Tissue culture studies on cereals. In: Reinert J, Bajaj YPS (eds) Applied and fundamental aspects of plant cell, tissue, and organ culture. Springer, Berlin Heidelberg New York, pp 144–159

Zaghmout OMF, Torello WA (1988) Enhanced regeneration in long-term callus cultures of red fescue by pretreatment with activated charcoal. HortScience 23: 615–616

Zaghmout OMF, Torello WA (1992a) Restoration of regeneration potential of long-term cultures of red fescue (*Festuca rubra* L.) by elevated sucrose levels. Plant Cell Rep 11: 142–145

Zaghmout OMF, Torello WA (1992b) Plant regeneration from callus and protoplasts of perennial ryegrass (*Lolium perenne* L.). J Plant Physiol 140: 101–105

Zhong H, Srinivasan C, Sticklen MB (1991) Plant regeneration via somatic embryogenesis in creeping bentgrass (*Agrostis palustris* Huds.). Plant Cell Rep 10: 453–456

Chapter 4
Anther Culture and Production of Haploids

4.1 Introduction

Haploid plants have the gametophytic number of chromosomes. They can be derived by several different ways. Haploids via twin seedlings can spontaneously occur in nature. Maternal haploids may arise from certain interspecific crosses followed by the elimination of chromosomes from the pollen donor during embryo development. By using in vitro culture techniques, haploid plants can also be regenerated from cultured anthers, microspores and ovaries. In this manner, both androgenetic (derived from microspores or immature pollen) and gynogenetic (derived from unfertilized egg cells) haploid tissue cultures and plants have been obtained in many different plant species. Gametes are the most convenient material for the reproducible production of haploids. Since normally in plants a larger number of haploid cells is concentrated in the anthers, which contain microspores or pollen, androgenesis theoretically represents the more efficient option. Therefore the techniques of anther culture and microspore culture have found wide application for the production of haploids, among others in several cereals and other grasses (Guha and Maheshwari 1964; Nitsch and Nitsch 1969; Nitsch 1977; Reinert and Bajaj 1977; Wenzel and Foroughi-Wehr 1984).

In anther and microspore cultures, haploid plants can be regenerated directly from the cultured immature pollen. In addition, the proliferation of these cells may result in callus formation which can then be induced to differentiate into, among others, haploid plants.

Doubled haploids, which theoretically are completely homozygous plants, can be generated from the haploid cells, cultures and regenerants after a spontaneous or chemically induced chromosome doubling. They are genetic equivalents of highly inbred lines if no mutations occur during any of the chromosome doubling and plant regeneration steps. Thus, anther culture provides means of recovering haploid and fertile homozygous doubled-haploid plants. The production and use of haploids and doubled haploids allows enhanced selection efficiency and speeds up crop improvement. It offers opportunities for identification and selection of recessive mutants and generates basic material for somatic hybridization and DNA marker mapping programs (Morrison and Evans 1988; Wenzel et al. 1995).

Table 4.1. Anther culture in forage and turf grasses

Plant species	Culture response	References
Festuca arundinacea (tall fescue)	Callus	Niizeki and Kita (1973)
	Callus	Niizeki (1977)
	Green plants	Kasperbauer et al. (1980)
	Green plants	Kasperbauer and Eizenga (1985)
Festuca pratensis (meadow fescue)	Globular embryoids	Zenkteler and Misiura (1974)
	Callus and albino plantlets	Rose et al. (1987a)
Lolium multiflorum (Italian ryegrass)	Callus and albino plantlets	Clapham (1971)
	Callus and green plant	Niizeki (1977)
	Callus and green plant	Nitzsche and Wenzel (1977)
	Chimeric plantlet and albino plantlets	Pagniez and Demarly (1979)
	Green plants and albino plantlets	Boppenmeier et al. (1989)
	Green plants and albino plantlets	Bante et al. (1990)
Lolium perenne (perennial ryegrass)	Callus	Niizeki (1977)
	Green plants and albino plantlets	Stanis and Butenko (1984)
	Green plants and albino plantlets	Olesen et al. (1988)
	Green plants and albino plantlets	Boppenmeier et al. (1989)
	Green plants and albino plantlets	Bante et al. (1990)
	Green plants and albino plantlets	Halberg et al. (1990)
	Green plants and albino plantlets	Hayward et al. (1990)
	Green plants and albino plantlets	Opsahl-Ferstad et al. (1994a,b)
	Green plants and albino plantlets	Madsen et al. (1995)
Lolium temulentum (annual ryegrass)	Callus and albino plantlets	Rose et al. (1987a)
Lolium multiflorum × *Festuca arundinacea* (*Festulolium* hybrid)	Green plant and albino plantlets	Nitzsche (1970)
	Green plants and albino plantlets	Nitzsche and Wenzel (1977)
Lolium multiflorum × *Festuca pratensis* (*Festulolium* hybrid)	Green plants and albino plantlets	Rose et al. (1987a)
Lolium perenne × *Festuca pratensis* (*Festulolium* hybrid)	Green plants and albino plantlets	Rose et al. (1987a)
Festuca pratensis × *Lolium multiflorum* (*Festulolium* hybrid)	Green plants and albino plantlets	Nitzsche and Wenzel (1977)
Agropyron cristatum (crested wheatgrass)	Callus	Marburger and Wang (1988)
Agropyron desertorum (crested wheatgrass)	Callus	Marburger and Wang (1988)
Bromus inermis (bromegrass)	Green plant	Saito et al. (1973)

Table 4.1. *Continued*

Plant species	Culture response	References
Dactylis glomerata (cocksfoot)	Callus and albino plantlet	Saito et al. (1973)
Pennisetum purpureum (napier grass)	Green plants	Haydu and Vasil (1981)
Pennisetum typhoides (pearl millet)	Green plants	Bui Dang Ha and Pernes (1982)
Phleum pratense (timothy)	Callus	Niizeki and Kita (1973)
	Callus	Niizeki (1977)
	Green plants and albino plantlets	Abdullah et al. (1994)
Setaria italica (foxtail millet)	Green plants	Ban et al. (1971)
Thinopyrum intermedium (intermediated wheatgrass)	Callus	Marburger and Wang (1988)
Thinopyrum ponticum (tall wheatgrass)	Callus and albino plantlets	Marburger and Wang (1988)

Anther culture technology may also prove valuable in genetic improvement of wind-pollinated, highly self-infertile forage and turf grasses such as tall fescue and perennial rygrass. In the case of these heterozygous outbreeders, there is a high probability that each gamete and gamete-derived haploid and doubled haploid plant may be genetically different from each other gamete or haploid and doubled haploid plant derived from another gamete. The evaluation in various environments of haploid plants derived from anther culture and cloned via tillers or in vitro tillering (see Sect. 2.2) would allow the identification of desirable genotypes under different field and controlled stress conditions. Chromosome doubling in the selected haploid plants should lead to fertile pure lines containing the traits selected during the evaluation of haploids (Kasperbauer 1990a,b). Furthermore, these fertile homozygous doubled haploids could be useful, as they are analogous to highly inbred lines, in crosses with other doubled-haploid plants to produce hybrids (Kaul 1990). The forced self-fertilization used in breeding programs to produce inbred lines strongly selects against the pronounced self-incompatibility naturally occurring in *Lolium* and *Festuca*. When these inbred lines are topcrossed or polycrossed, a certain proportion of self-pollination may occur, thus leading to wrong estimations of breeding values (Utz and Oettler 1978). The recovery of homozygous plants from anther culture may overcome this problem, since no selection against self-incompatibility is expected in this case (Creemers-Molenaar and Beerepoot 1992). In addition, homozygous plants from anther culture-derived haploids might prove useful in producing multiline synthetics by crossing with other doubled haploids or with conventionally generated lines.

The following sections review published studies devoted to the production of androgenetic haploids from microspores in forage and turf grasses (Table 4.1). Special attention is given to corresponding reports on culture of excised anthers and anther-panicle nurse culture in fescues and ryegrasses.

4.2 Anther Culture and Haploids in *Festuca*

Although first attempts to produce androgenetic haploids in grasses belonging to the *Festuca-Lolium* complex were undertaken early on as in cereals (Nitzsche 1970; Clapham 1971), significantly less efforts have been committed to anther culture studies in ryegrasses and fescues over the last two decades (Table 4.1). First results on anther culture in fescues were reported for *F. arundinacea* and *F. pratensis* (Niizeki and Kita 1973; Zenkteler and Misiura 1974; Niizeki 1977). However, the recovery of haploid androgenetic fescue plants was not achieved in any of these early cases. In *F. arundinacea*, according to Miller (1963) and Nitsch and Nitsch (1969), out of 150 and 99 anthers cultured on M- and N&N-derived basal media, respectively, only one anther-derived callus was obtained on N&N medium supplemented with 10mg/l indole-acetic acid (IAA) and 10mg/l kinetin (Niizeki and Kita 1973). Evidence from microscopic observations of the cultured anthers indicated that the callus obtained might have been derived from a uninucleate pollen grain after some initial embryogenic development. In the same study, no development was observed for 90 anthers plated on M medium in *F. pratensis* (Niizeki and Kita 1973). No improvement in anther culture response was found in analogous studies involving 806 anthers plated from six different cultivars of tall fescue and 180 cultured anthers from two cultivars of meadow fescue (Niizeki 1977). Zenkteler and Misiura (1974) plated 3104 anthers of *F. pratensis* on media derived from basal LS medium (Linsmaier and Skoog 1965) and N&N medium (Nitsch and Nitsch 1969). Globular embryoids developing from microspores were detected in 0.5% of the meadow fescue anthers cultured on the basal LS-medium supplemented with 2mg/l kinetin, 1mg/l IAA and 4% sucrose, but the embryoids failed to develop further and no haploid callus formation was observed (Zenkteler and Misiura 1974).

Important progress in the recovery of androgenetic plants of *F. arundinacea* was made by Kasperbauer et al. (1980) with the adoption of the anther-panicle nurse culture method. Attempts to obtain haploid callus or plants from more than 2000 excised anthers of tall fescue containing microspores at the uninucleate stage failed after testing various culture media and preculture temperature regimes as well as several cultivars and ecotypes from two different growing seasons (Kasperbauer et al. 1980; Kasperbauer 1990a).

However, polyhaploids ($2n = 3x = 21$), i.e. haploids of the polyploid (hexaploid) tall fescue, could only be recovered when anthers attached to a segment of panicle tissue to exert a nurse effect were cultured (Kasperbauer et al. 1980). After 7 weeks in culture, several green plantlets were obtained from the cultured 2- to 3-cm anther-panicle explants obtained from one endophyte-free plant, a single genotype of the cultivar Kentucky 31. Only green plantlets were regenerated and 23 adult plants were established in soil. The chromosome counts performed in shoot tips revealed the polyhaploid nature ($2n = 3x = 21$) of all but one ($2n = 39$) of these plants (Kasperbauer et al. 1980). The cytologically verified polyhaploid plants derived from anther-panicle nurse culture and clonally multiplied by tillers showed differences in phenotypic and forage quality traits when grown in replicated field plots (Kasperbauer and Eizenga 1985; Kasperbauer 1990a). These 22 polyhaploid tall fescue lines were evaluated for winter hardiness, growth and development. In addition, sugar, total nitrogen, moisture, dry matter digestibility, neutral detergent fiber and phosphorus levels were analyzed to estimate forage quality in summer and fall (Kasperbauer and Eizenga 1985). Pronounced phenotypic differences existed among the various polyhaploid lines, ranging from early flowering with many panicles and little leaf growth to sparse flowering with abundant leaf growth (Kasperbauer 1990a). In contrast, field growth from 15 ramets planted per polyhaploid line was very uniform within each polyhaploid line evaluated. Significant differences among the polyhaploid lines occurred also for the chemical analyses of selected forage characterisitics, e.g. sugar, neutral detergent fiber and moisture ranged from 5.0 to 7.8%, 48 to 59%, and 63 to 72%, respectively, for the summer harvest of field-grown samples (Kasperbauer and Eizenga 1985). The differences for the measured traits observed among the various polyhaploid lines and the uniformity and high stability demonstrated upon repeated cloning via tillers within each polyhaploid line support the concept that these lines originated from different microspores from a heterozygous donor plant and were genetically different. Thus, the identification of different genotypes among tall fescue polyhaploid plants was possible (Kasperbauer and Eizenga 1985; Kasperbauer 1990a). Meiotic chromosome pairing relationships and female fertility were examined in these polyhaploids of tall fescue (Eizenga and Kasperbauer 1985). They had smaller panicles and shorter, narrower leaves than the tall fescue plant from which they were derived. A low level of female fertility could be demonstrated by the production of eupoloid, monosomic and double-monosomic progeny from the open-pollinated polyhaploid plants in this study. Chromosome associations at metaphase I in six tall fescue polyhaploids ranged from 15.98 to 20.40 univalents, 0.30 to 2.56 bivalents, and 0.00 to 0.18 trivalents (Eizenga and Kasperbauer 1985). The observed variation in the level of chromosome pairing is understandable based on the different genotypes of the individual polyhaploids derived from a heterozygous tall fescue plant. In contrast, the low chromosome pairing observed by Eizenga and Kasperbauer (1985) is difficult to explain considering that the pairing control mechanism of tall fescue breaks down in its haploid complement in all hybrids of *Lolium-Festuca* species (Jauhar 1993).

The tall fescue polyhaploids obtained by Kasperbauer et al. (1980) were largely sterile and no spontaneously doubled haploids were detected among the tiller-propagated plants of any of the polyhaploid lines tested. Therefore, a protocol for induced chromosome doubling was worked out to obtain fertile doubled haploids in *F. arundinacea* (Kasperbauer and Eizenga 1985; Kasperbauer 1990b). Since conventional colchicine treatment (Buckner et al. 1961) failed to produce doubled haploids from tall fescue polyhaploids, Kasperbauer and Eizenga (1985) developed a procedure for the regeneration of doubled-haploid shoots from somatic tissue excised from cytologically verified polyhaploids. Explants from the base of rapidly elongating peduncles and internodes from six verified tall fescue polyhaploids were plated on LS basal medium supplemented with 2 mg/l 2,4-D and no cytokinin. Within 4–5 weeks the calli formed were excised, subcultured once and "aged" for 15 weeks until they began to senesce. "Aged" calli were transferred onto regeneration medium containing 0.25 mg/l 2,4-D, and shoots formed within 4 weeks could be transferred to rooting medium deprived of auxin and cytokinin. Out of 60 plants regenerated using this procedure, 37 were haploid ($2n = 3x = 21$), 14 were doubled haploid ($2n = 6x = 42$), and the remaining regenerants showed somatic chromosome numbers varying between 38 and 84. One or more doubled-haploid tall fescue plants were regenerated from each of the six different polyhaploid lines included in this study (Kasperbauer and Eizenga 1985). Few albino or abnormal plantlets were observed among the in vitro regenerants obtained from these callus cultures, thus providing evidence for the occurrence of some somaclonal variation (see Chap. 3).

In general terms, however, the doubled-haploid plants had the same markers, e.g. leaf angle and shape, as the respective haploid source plant. Some of the haploids, all doubled haploids, and one 84-chromosome regenerated plant were cloned via tillering and used for further evaluations. Particularly the series of cloned 21-, 42- and 84-chromosome plants derived from one of the original polyhaploids obtained by anther-panicle nurse culture offered the opportunity to study ploidy effects on various plant traits and forage characteristics, e.g. winter hardiness, succulence, crude protein, in vitro dry matter digestibility and potassium/calcium plus magnesium ratio (Kasperbauer et al. 1987; Kasperbauer 1990b). The doubled haploid tall fescue plants grown under field conditions were fertile and set seeds with germination frequencies comparable to the corresponding controls (Kasperbauer 1990b). Results of the cytological and agronomic comparative evaluation of progenies from the regenerated doubled-haploid tall fescue plants with conventionally derived plants have not yet been reported. This evaluation, however, would provide further valuable information to assess the feasibility of incorporating these materials together with conventionally derived ones in tall fescue breeding programs.

A few basic studies on the methods and cytology of anther culture are also available for *F. pratensis* and intergeneric *Festulolium* hybrids (Table 4.1; Nitzsche 1970; Nitzsche and Wenzel 1977; Rose et al. 1987a, b). Rose et al.

(1987a) evaluated the influence of different durations (7–42 days) of panicle pretreatments at 4 and 7 °C, different media with various concentrations of 2,4-D and kinetin with and without added charcoal, as well as different incubation temperature and pretreatment/incubation light regimes on yield of microspore-derived callus in *F. pratensis* and the two hybrids *L. multiflorum* × *F. pratensis* cv. Elmet and *L. perenne* × *F. pratensis* cv. Prior. The pathways of pollen development in vitro were investigated in cultured anthers of *F. pratensis* and the *L. multiflorum* × *F. pratensis* hybrid (Rose et al. 1987b). This cytological study showed that pollen callus development was predominantly from the vegetative cell and that some variation in behaviour patterns from the generative cell were evident. In *F. pratensis* the generative cell preferentially divided, while in the *Festulolium* hybrid it became detached from the pollen wall and remained quiescent (Rose et al. 1987b).

Anther response and callus yield varied with the duration and temperature of the pretreatments. Maximum anther response was achieved after 42 days at 4 °C in *F. pratensis*. Similarly, pretreatments at 4 °C for more than 14 days gave the best anther response in the hybrid *L. multiflorum* × *F. pratensis*. In both cases, up to 10% of the plated anthers formed androgenetic callus. In contrast, for the hybrid *L. perenne* × *F. pratensis*, anther culture reponse tended to decline with increasing duration of pretreatments at both temperatures, and a maximum response was apparent after 7 days. The frequency of cultured anthers forming androgenetic callus varied in the range of 10–30% for this *Festulolium* hybrid (Rose et al. 1987a). Maximum callus yields from cultured anthers of meadow fescue occurred after 28 and 35 days at 4 °C, while no effect of duration and temperature of panicle pretreatments on androgenetic callus yield were found for the *L. multiflorum* × *F. pratensis* hybrid. Similarly, callus yield seemed to be hardly affected by the duration of the cold pretreatment of panicles from the hybrid *L. perenne* × *F. pratensis* (Rose et al. 1987a).

For *F. pratensis*, no significant differences in anther response or callus yield were detected in the evaluation of four media with various concentrations of 2,4-D and kinetin (Rose et al. 1987a). The addition of charcoal to the media was found to inhibit androgenetic callus growth. Differentiation of androgenetic calli in meadow fescue was very poor in all cases, and only 4 albino plantlets regenerated out of 1802 calli obtained (Rose et al. 1987a). In contrast, the regeneration of both green and albino plantlets was possible from androgenetic calli of the hybrids *L. multiflorum* × *F. pratensis* and *L. perenne* × *F. pratensis* (Rose et al. 1987a). Out of 1356 calli obtained from cultured anthers of different panicles in the hybrid *L. multiflorum* × *F. pratensis*, 14 calli differentiated green, albino or yellow shoots, and 4 of them gave green plants. For the *L. perenne* × *F. pratensis* hybrid, 6868 calli were produced, 191 showed differentiation of shoots or roots, and 22 of them led to the recovery of green plants. While 32.6% of the calli arose from panicles pretreated at 7 °C, only 3 of the 22 green plants were produced with this treatment. Furthermore, significant differences in both anther response and callus growth were found in the evaluation of individual genotypes of the *L. perenne* × *F. pratensis* hybrid (Rose et al. 1987a).

The differentiation of only albino plantlets from androgenetic calli in the diploid *F. pratensis* and also in the diploid *L. temulentum* (see Sect. 4.3) observed in this study (Rose et al. 1987a) contrasts with the recovery of green, albino, chimeric and yellow shoots and plants in the amphidiploid *L. multiflorum* × *F. pratensis* and *L. perenne* × *F. pratensis* hybrids. Although a direct comparison is not warranted in this case, the differences observed in the occurrence of albinism could be explained by the genotype dependence of this phenomenon. In addition, it can be speculated that the ploidy level of the donor plant might have further contributed to these differences, since recessive mutations are more likely to be phenotypically revealed in regenerants from haploid pollen of the diploid species (*L. temulentum* and *F. pratensis*) than in regenerants from diploid pollen produced in the tetraploid *Festulolium* hybrids. The accompanying cytological study (Rose et al. 1987b) did not provide any evidence of pollen callus development in the *L. multiflorum* × *F. pratensis* hybrid occurring via a pathway not operative in *L. temulentum* or *F. pratensis* that could have explained the production of green androgenetic plants in the *Festulolium* hybrid but none in the diploid species.

Nitzsche (1970) reported the production of one androgenetic callus, which differentiated one green plant and three albino plantlets, out of 488 anthers cultured from a 49-chromosome *L. multiflorum* (4x) × *F. arundinacea* (12x) hybrid. The green plant recovered showed 25 chromosomes. Results from subsequent anther culture experiments with *Festuca* × *Lolium* hybrids and their reciprocal hybrids have been briefly described by Nitzsche and Wenzel (1977). Out of 11 androgenetic calli obtained from 824 plated anthers of an octoploid (2n = 8x) *L. multiflorum* (4x) × *F. arundinacea* (12x) hybrid, 4 green plants were regenerated. Four calli regenerating albino plantlets were obtained from 17997 anthers from a diploid *F. pratensis* (2x) × *L. multiflorum* (2x) hybrid cultured on modified Blaydes (1966) medium. For the tetraploid *F. pratensis* (4x) × *L. multiflorum* (4x) hybrid, 46 androgenetic calli were recovered out of 29220 plated anthers and 6 of them regenerated green dwarf plants (Nitzsche and Wenzel 1977). Thus, the most interesting feature of the anther culture experiments with *Festuca* × *Lolium* reciprocal hybrids is the fact that androgenesis, leading finally to the recovery of polyhaploid green plants, can be induced even in sterile hybrids.

4.3 Anther Culture and Haploids in *Lolium*

Anther culture with *Lolium* species had its origin in the pioneering work by Clapham (1971). Callus development from uninucleate microspores was achieved in up to 20% of anthers excised from *L. multiflorum* when they were cultured on basal White or LS media supplemented with 12% sucrose, 15%

coconut milk and 1 mg/l NAA (Clapham 1971). Overall regeneration frequencies of 3% from the plated anthers were reported. However, only albino plantlets were recovered in this early study.

Few reports devoted to anther culture in ryegrasses over the following 10 years did not reflect any significant improvement (Niizeki 1977; Nitzsche and Wenzel 1977; Pagniez and Demarly 1979). From 520 anthers cultured in an autotetraploid cultivar of *L. multiflorum,* a single haploid ($2n = 2x = 14$) plant showing 92% pollen fertility regenerated out of 7 pollen-calli obtained by Niizeki (1977). Only two calli and no plants were recovered from *L. perenne* out of 150 anthers cultured on LS medium containing 12% sucrose in the same study (Niizeki 1977). Nitzsche and Wenzel (1977) described the recovery of 36 calli and a single haploid green plant from 7650 anthers cultured in tetraploid *L. multiflorum.* Mainly androgenetic calli and embryoids were obtained from cultured anthers in one diploid and one tetraploid synthetic cultivar of *L. multiflorum* by Pagniez and Demarly (1979). A maximal response frequency of 32% of anthers producing androgenetic structures was achieved under optimum culture-medium (with 2,4-D and high sucrose concentration) and temperature (21–22 °C) conditions, and a marked genotype-effect in androgenetic response was also found (Pagniez and Demarly 1979). Besides many albino plantlets, only one chimeric plantlet consisting of albino and green shoots was obtained. The chimeric in vitro regenerant was tetraploid as the corresponding Italian ryegrass plant used as anther donor (Pagniez and Demarly 1979).

Both observations, genotype dependence in anther culture response and recovery of non-haploid plants, have been confirmed in *L. perenne* (Stanis and Butenko 1984). Out of 9000 cultured anthers excised from four cultivars and two ecotypes of perennial ryegrass, 676 calli were obtained. These led finally to the recovery of 530 albino and 100 green plants (Stanis and Butenko 1984). Depending on the plant material used, the number of calli produced per anther varied between 1.3 and 3.9. Similarly, the number of regenerants per androgenetic callus varied in the range of 1.7 to 3.0 (Stanis and Butenko 1984). Genotypic differences were thus inferred for callus induction in anther culture, for regeneration capacity and green : albino ratio from induced calli, as well as for the ploidy level of the perennial ryegrass plants recovered. No interdependence between callus formation rates and regeneration ability was found. Besides haploids, plants of other ploidy levels (diploid, triploid, tetraploid) and, in addition, one mixoploid plant were obtained from these anther cultures in *L. perenne* (Stanis and Butenko 1984).

Analogous information on the methodology and cytology of anther culture as described for *F. pratensis* and two intergeneric *Festulolium* hybrids in Sect. 4.2 is available for the annual ryegrass, *L. temulentum* (Rose et al. 1987a, b). Both anther response and callus yield were found to increase with increased duration of spike pretreatment at 7 °C for up to 28 days and then declined. Although 5115 androgenetic calli were obtained in these experiments, the inclusion of activated charcoal in corresponding media strongly inhibited root or shoot differentiation (Rose et al. 1987a). In self-pollinating *L. temulentum*, the evaluation of different media constituents and light and subculture regimes allowed for up to 30.7% of the cultured anthers to produce androgenetic calli with up to 2.4 calli generated per anther (Rose et al. 1987a). However, differentiation from the androgenetic calli remained poor, and only 25 albino plantlets could be recovered out of 1349 pollen-calli obtained (Rose et al. 1987a). Development from the vegetative cell and degeneration of the generative cell attached to the pollen wall were found to be the predominant pathway of pollen callus development in anther cultures of *L. temulentum* (Rose et al. 1987b).

In anther culture of ryegrasses, considerable progress has been made following the contribution by Olesen et al. (1988) using media and methods adapted from wheat anther culture. A total of 65745 anthers containing microspores in late uninucleate stage were cultured onto PII medium (Wang and Hu 1984) supplemented with 2.5% potato extract and solidified with 0.3% gelrite from 20 diploid and 10 tetraploid clones from 7 and 3 varieties of *L. perenne*, respectively (Olesen et al. 1988). Androgenetic calli or embryoids were formed in 14.7% of the cultured anthers, and 12155 albino and 463 green plantlets were obtained out of 41730 pollen-derived embryoids and calli transferred onto regeneration medium 190-2 (Wang and Hu 1984). Of the 463 green regenerants, 226 plants were established in soil. Androgenetic structures were formed for all 30 clones tested. However, marked genotypic differences were found, varying in the range of 0.3–164 per 100 cultured anthers (Olesen et al. 1988). Albino plantlets were regenerated from all but one clone, while green plants were recovered from 12 different clones – 5 diploid and 7 tetraploid – out of the 30 tested. On average 19.2 plants were obtained per 100 anthers cultured, but green regenerants were still rare for most clones. For two clones, however, an overall average of 2.3 and 3.8 green plants per 100 cultured anthers was found (Olesen et al. 1988). Furthermore, estimates of variance components from replicates of greenhouse and field-grown donor plants showed that genotypes accounted for about 73% of the total variation observed in yield of androgenetic structures. Monoploid ($2n = x = 7$) and diploid ($2n = 2x = 14$) green plants were found in 21 and 29 cases, respectively, among regenerants from cultured anthers of diploid donor plants; while 14 diploid ($2n = 2x = 14$) and 4 tetraploid ($2n = 4x = 28$) plants were recovered from anther culture of tetraploid donor plants (Olesen et al. 1988). The isozyme analysis performed for the phosphoglucoisomerase polymorphic locus 2 (*Pgi 2*) and glutamate-oxaloacetate transaminase polymorphic loci 2 and 3 (*Got 2* and *Got 3*) in 25 recovered diploid plants, confirmed homozygosity at these loci, while the corresponding anther donor plants were heterozygous in the loci investigated. Thus it could be demonstrated that for diploid donor perennial ryegrass, 50–60% of the regenerants from anther culture were doubled haploids (Olesen et al. 1988). This frequency of spontaneous chromosome doubling in regenerants from anther culture using diploid *L. perenne* donor plants has been confirmed (Hayward et al. 1990).

The results obtained by Olesen et al. (1988) with *L. perenne* indicated that: (1) anther culture-responsive genotypes can be found in different perennial ryegrass cultivars; (2) the genotype of the corresponding donor plants largely determines the pollen-callus induction response; (3) the use of single responsive genotypes rather than cultivars would be recommendable for further optimization of anther culture protocols; (4) anther culture can be used for the recovery of homozygously diploidized perennial ryegrass plants; and (5) the generation of genotypes with superior anther culture response after high-responding genotypes have been intercrossed should be testable as a first step towards breeding for anther culture response in addition to agronomic performance through hybridization and selection.

These conclusions have been confirmed and in part also extended to *L. multiflorum* (Fig. 4.1; Boppenmeier et al. 1989; Bante et al. 1990; Halberg et al. 1990; Hayward et al. 1990; Madsen et al. 1993, 1995; Opsahl-Ferstad et al. 1994a, b). Using modifications of barley anther culture techniques (Foroughi-Wehr et al. 1976), Boppenmeier et al. (1989) succeeded in regenerating green and albino plants from cultured anthers in *L. multiflorum* ssp. *gaudini* and *L. perenne*. Out of 1872 cultured anthers in one cultivar of *L. multiflorum*, 40 responding anthers finally regenerating 18 albino and 8 green plants were recovered. Analogous experiments involving 44 clones of perennial ryegrass resistant to barley yellow dwarf virus (BYDV) and 5 different *L. perenne* cultivars were performed. Of 61716 cultured anthers, 1352 formed androgenetic structures which regenerated into 811 albino and 452 green plants (Boppenmeier et al. 1989). Significant genotypic differences in pollen-callus induction and plant regeneration could be demonstrated in perennial ryegrass. However, a high frequency of plant formation was not always associated with maximum green plant regeneration. In addition, a high-responding genotype capable of producing up to 20 green plants per 100 cultured anthers was selected. A significant genotypic variation in response to temperature treatment (22 and 30 °C) was also apparent (Boppenmeier et al. 1989). Of the 130000 anthers plated from the 44 BYDV-resistant clones, 34 green plants suitable for initial genetic tests on the inheritance of BYDV resistance in *Lolium* were obtained from 8 selected clones. The determination of ploidy level in 401 adult plants out of 452 regenerants from anther culture established in soil revealed 70% of them to be haploids ($2n = x = 7$) and 29% to have spontaneously diploidized ($2n = 2x = 14$). One triploid ($2n = 3x = 21$) and one tetraploid ($2n = 4x = 28$) plant were also recovered (Boppenmeier et al. 1989). The identification of a high-responsive genotype from one *L. perenne* cultivar as well as of responsive BYDV-resistant clones suggested the possibility of further optimizing regeneration frequencies of haploids and doubled haploids and using haploids in breeding to combine BYDV resistance with other traits of agronomic interest.

Optimization of anther culture response in *L. multiflorum* and *L. perenne* has been reported by modifying corresponding culture conditions (Bante et al. 1990; Opsahl-Ferstad et al. 1994a). Based on 1200 *L. perenne* and 4000

A
B
C
D
E
F
G

L. multiflorum anthers plated from different varieties, a cold pretreatment of splikelets at 4 °C for 2–3 weeks in darkness was found to enhance four- to fivefold the frequency of anthers producing androgenetic calli or embryoids (Bante et al. 1990). Out of 52752 cultured anthers of different perennial ryegrass genotypes for evaluating effects on anther culture response of a cold pretreatment at 4 °C, Opsahl-Ferstad et al. (1994a) were able to recover 46173 androgenetic structures and regenerated 5466 albino and 401 green plants. These authors found that cold treatment significantly increased the number of androgenetic embryoids, the number and proportion of albino plants formed, but not the production of green plants. No significant effect of the duration of cold treatment was found (Opsahl-Ferstad et al. 1994a).

A similar (four- to fivefold) increase in responding anthers was achieved upon incubation of 3000 anthers each in *L. perenne* and *L. multiflorum* at elevated (2%) CO_2 instead of atmospheric CO_2 concentration (Bante et al. 1990). A tenfold reduction of the ammonium nitrate concentration in the induction medium to 2 m*M* allowed doubling of the number of responding anthers in both ryegrasses. The substitution of sucrose, which normally is used in *Lolium* anther culture at concentrations of 12 (Clapham 1971) or 9% (Olesen et al. 1988), by maltose, trehalose or maltotriose increased the frequency of anthers producing androgenetic calli by a factor of 2–6, leading to values of up to 18% in *L. perenne* and up to 36% in *L. multiflorum* (Bante et al. 1990). Maltose, maltotriose and trehalose appeared to not only increase the anther response but also the overall culture efficiency of the 1250 *L. perenne* and 1850 *L. multiflorum* anthers tested in this study. Both green and albino plantlets were recovered from cultured anthers in Italian and perennial ryegrass. Again, a clear genotype effect on anther culture response and on the ratio of green to albino regenerants was found. Other media constituents, namely different basal media, gelling agents, and addition of activated charcoal in a double layer-design and of casein hydrolysate, have been tested by Opsahl-Ferstad et al. (1994a) for their effect on androgenetic response in different genotypes of *L. perenne*. Significant differences were found between media, genotypes and medium-genotype interactions (Opsahl-Ferstad et al. 1994a). However, the authors concluded that all commonly used anther culture media in perennial ryegrass, namely PII, 190-2 and LS-3 (Wang and Hu

Fig. 4.1A–G. Anther culture and regeneration of dihaploids in *Lolium multiflorum* Lam. **A** Newly plated anthers of *L. multiflorum*. **B** Formation of androgenetic embryoids and calli from anthers float-cultured on liquid medium for 6 weeks. **C** Androgenetic embryoids from anthers cultured on solidified medium for 6 weeks. **D** Green and albino regenerants from androgenetic calli obtained 9 weeks after plating anthers of *L. multiflorum*. **E** Albino regenerant from cultured anthers of *L. multiflorum*. **F** In vitro-growing plantlets regenerated from androgenetic calli 3 months after initiation of corresponding cultures. **G** six-month old flowering fertile dihaploid Italian ryegrass plant from anther culture. (Courtesy of U.K. Posselt, Saatzuchtanstalt, Universität Hohenheim, Stuttgart, Germany)

1984; Olesen et al. 1988; Bante et al. 1990), were equivalent with regard to production of green plants. Thus medium PII could be avoided so as to reduce the uncontrollable variation associated with the use of potato extract (Opsahl-Ferstad et al. 1994a).

Bante et al. (1990) used flow cytometric analysis to evaluate the ploidy level of 234 anther-derived Italian and perennial ryegrass plants and found 30% of them to have a gametic chromosome number, while the others were mainly diploid and tetraploid. In addition, a few pentaploid, hexaploid, octoploid, aneuploid and mixoploid plants were also obtained. Of 44 anther culture-derived perennial ryegrass plants from diploid anther donors, 24 diploids (55%), 12 haploids (27%) and 7 polyploids were found. In the case of the diploid Italian ryegrass anther donor, of 67 anther-derived plants evaluated, 49 diploids (73%), 6 haploids (9%) and 12 polyploids (18%) were detected (Bante et al. 1990). The microspore origin of these diploids could be confirmed by PGI 2 isozyme analysis, and thus differences in the frequency of spontaneous diploidization were apparent between the *L. multiflorum* and *L. perenne* genotypes considered. While plants derived from anther culture of tetraploid Italian ryegrass consisted of about one half diploids (42 of 89) and one half tetraploids (44 of 89) and a few plants of higher ploidy, a slightly different ploidy level distribution was found for regenerants from cultured anthers excised from tetraploid perennial ryegrass donors (Bante et al. 1990). These relatively high frequencies (50–70%) of spontaneous diploidization from diploid ryegrass anther donors (Olesen et al. 1988; Bante et al. 1990; Hayward et al. 1990) clearly differ from the relatively low value (less than 30%) reported by Boppenmeier et al. (1989).

An examination of the segregation of isozyme markers in a set of green and albino microspore-derived progenies of two plants of *L. perenne* has been reported as a first assessment of type and range of genetical variation produced by anther culture (Hayward et al. 1990). Both parents were heterozygous at the *Pgi 2* and *Got 2* loci, whilst all progeny were homozygous for one or the other allelic forms, thus confirming their microspore origin. Spontaneous doubling to the diploid level was found in 48% of the green regenerants. The overall pooled data for both loci did not deviate from the 1:1 ratio expected. However, considerable heterogeneity between families and between green and albino plants was found (Hayward et al. 1990). Selection of one allele of each isozyme from the one donor clone and selection of the other allele from the second donor clone were consistently found. The predominance of specific alleles in the different progeny sets may be accounted for by possible linked loci affecting anther culture response or by linked lethal genes. The selection taking place, however, will restrict the spectrum of genetic variation among the corresponding doubled haploids. The implications of this selection for breeding purposes will depend on its nature, e.g. whether major genes affect androgenetic capacity or gametophytic selection against linked lethal or semi-lethal genetic factors is active. Further studies are required to clarify this point.

Halberg et al. (1990) have convincingly shown that superior anther culture responding genotypes of diploid *L. perenne* can be constructed by sexual crossing of three previously identified responsive primary materials (Olesen et al. 1988). Of 41432 anthers cultured during this study, 28502 androgenetic embryoids were formed, from which 9830 albino and 1390 green plants differentiated. On overall average, 68.8 embryoids and 3.4 green plants were recovered per 100 cultured anthers. Significant genotype effects on androgenetic embryoid formation, embryoid differentiation into plantlets, percentage of green plant regeneration and recovery of green plants were found (Halberg et al. 1990). Among 55 hybrid clones, 6 hybrids were identified with an anther culture response level superior to their parents. While the parents produced one green plant at most per 100 cultured anthers, these superior clones were able to generate 11–59 green plants per 100 plated anthers, particularly as a result of enhanced production of androgenetic embryoids and an increased percentage of green plant differentiation (Halberg et al. 1990). Provided the anther culture response were highly heritable in crosses with non-responsive materials of good agronomic performance, these superior clones might then be used as inducers of anther culture response by crossing with the non-responding breeding materials, yet requiring a few cycles of backcrossing.

The self-fertility of chromosome-doubled haploid lines of perennial ryegrass and their first and second generation selfed offspring have been studied by Madsen et al. (1993). From 510 plants tested, 3910 spikes were harvested, finally leading to 3084 plants. As expected for chromosome-doubled haploids of a self-incompatible species, an overall low seed set ranging between 0 and 0.4 seeds per spike was found. However, significant genetic variation was observed for self-fertility within three of six lines considered (Madsen et al. 1993). The increased self-fertility observed in a small number of clones of first and second generation selfing offspring suggests a selection during selfing of rare pollen carrying mutations in the self-incompatibility system. This would in turn lead to a general breakdown of self-incompatibility after repeated selfing of doubled haploids in perennial ryegrass and would open up opportunities for their use in future breeding of hybrid varieties based on cytoplasmic male sterility.

Studies on the genetic control of anther culture response in *L. perenne* have become available (Opsahl-Ferstad et al. 1994b; Madsen et al. 1995). Opsahl-Ferstad et al. (1994b) studied androgenesis in 90 perennial ryegrass genotypes, of which 82% were able to produce androgenetic embryoids, 71% produced albino plantlets and 17% regenerated green plants. High heritabilities of all characters controlling androgenetic response, except for percentage of green plant production, were found. Genotypic correlations between production of androgenetic embryoids (embryoids/100 anthers), total regeneration (plants/100 embryoids), and albino plant production (albino plantlets/100 anthers) were highly positive, whereas genotypic correlations between total regeneration and the green plants produced (green plants/100 anthers) were low. Opsahl-Ferstad et al. (1994b) evaluated anther culture response in F_1 plants from six possible reciprocal crosses made for three selected parental genotypes showing significantly different androgenetic response in the primary screening. Comparison of parent and progeny means showed that the androgenetic responses in the progeny were intermediate or at the level of the lower parent. However, in F_1 progenies from a cross between two relatively high-level producers of green anther-derived plants, an average superiority over both parents with regard to total plant regeneration and green plant production was found (Opsahl-Ferstad et al. 1994b). Thus, the androgenetic response in perennial ryegrass may be characterized by both qualitative and quantitative inheritance, with few loci being involved.

The inheritance of anther culture response in *L. perenne* has been independently studied by Madsen et al. (1995) for F_1 progenies obtained from crosses between 11 inducer clones producing an average of 10.7 green plants per 100 cultured anthers and 10 clones selected among breeding material for agronomic performance which regenerated only 0.3 green plants/100 anthers. On average, 2.2 green plants per 100 plated anthers were produced by the hybrid populations, and 75 of the 99 hybrid families studied produced green plants. However, the response rate was found to vary considerably between genotypes. This sevenfold increase in response over the breeding materials was mainly brought about by an enhancement of green plant regeneration to 4.7% compared to an average value of 0.5% for the breeding clones (Madsen et al. 1995). In agreement with the results reported by Opsahl-Ferstad et al. (1994b), the genetics behind the anther culture response in *L. perenne* seem to involve a few major genes and the trait seems to be highly recessive (Madsen et al. 1995). Furthermore, it could be shown that it is possible to transfer the capacity for green plant regeneration in anther culture to progeny from crosses between selected highly responsive inducer lines and breeding material of *L. perenne*, since all breeding clones in combination with at least one of the inducer clones gave progeny producing green plants (Madsen et al. 1995).

4.4 Anther Culture in Other Grasses

Over the last two decades, significant efforts have been made to improve doubled-haploid production via anther culture in some cereals. This method is increasingly applied to practical cereal breeding programs (Genovesi 1990; Picard et al. 1990; Xu 1990; Zhu and Pan 1990; Foroughi-Wehr and Wenzel 1993), and new cultivars have been developed by application of this methodology in wheat (De Buyser et al. 1987) and rice (Huang et al. 1988). Reports on the formation of androgenetic calli and plants from anther culture in different grasses besides the *Festuca-Lolium* complex also date back to the

1970s (Ban et al. 1971; Niizeki and Kita 1973; Saito et al. 1973). However, little progress has been made since then.

The recovery of haploid and diploid plants, some of which show relatively high pollen fertility, from anther culture has been described for *Setaria italica* (Ban et al. 1971). Saito et al. (1973) have reported low-frequency erratic production of androgenetic calli from anther culture in *Dactylis glomerata* and octoploid (2n = 56) *Bromus inermis*, finally leading to the differentiation of one albino plantlet and one tetraploid (2n = 28) green plant, respectively (Table 4.1). In *Phleum pratense*, formation of 8 androgenetic calli from 747 cultured anthers has been reported (Niizeki and Kita 1973; Niizeki 1977). The same authors have failed to obtain androgenetic calli from 249 and 365 cultured anthers from different cultivars of *B. inermis* and *D. glomerata*, respectively (Niizeki 1977). Significant improvement in anther culture of *P. pratense* has been achieved by using techniques developed for wheat and ryegrasses (Abdullah et al. 1994). A total of 62 regenerants, 49 green and 13 albino, were obtained from 65800 anthers cultured from 165 different timothy clones. Regeneration via multiple secondary embryogenesis was observed. Of the green androgenetic plants obtained from hexaploid *P. pratense*, 22% and 72% were triploid and hexaploid, respectively (Abdullah et al. 1994). Although the inheritance of the isozyme banding patterns obtained was unknown, evidence was obtained along these lines that the regenerated hexaploid plants were derived from triploids through spontaneous chromosome doubling (Abdullah et al. 1994).

Over 90 diploid (2n = 4x = 28) plants from embryogenic callus apparently obtained from somatic tissues of cultured anthers have been regenerated through somatic embryogenesis in *Pennisetum purpureum* (Haydu and Vasil 1981). Bui Dang Ha and Pernes (1982) regenerated plants from androgenetic embryoids and calli obtained in microspore cultures of *Pennisetum typhoides*. The corresponding chromosome counts performed and isozyme patterns studied for esterases and peroxidases in a set of the plants recovered revealed: (1) most plants to be diploid, with some being haploid, aneuploid or mixoploid; and (2) disturbed segregations of the biochemical markers with a preponderance of a single allelic form regenerating for the isozyme locus peroxidase *P5* (Bui Dang Ha and Pernes 1982).

Several genotypes of the Triticeae perennial forage grasses *Agropyron desertorum*, *A. fragile*, *A. cristatum*, *Thinopyrum intermedium* and *T. ponticum* have been evaluated for their anther culture response (Marburger and Wang 1988). Androgenetic calli were obtained for some accessions of *A. desertorum*, *A. cristatum* and *T. intermedium*, but failed to differentiate into plantlets. No response in any induction media tested was found in cultured anthers of *A. fragile*. In the case of anther cultures of *T. ponticum*, highest callus induction and plantlet production frequencies of up to 25 and 46% were observed, respectively. However, all plantlets regenerated from anther cultures in tall wheatgrass were albino (Marburger and Wang 1988).

4.5 Summary and Conclusions

Anther culture aimed at the production of androgenetic haploids and doubled haploids has been reported for representative fescues and ryegrasses over the last two decades. The development of the corresponding methods is still largely in its infancy. If perfected, however, haploid techniques are expected to be increasingly attractive in forage grass breeding in the near future. Particularly for perennial ryegrass, prospects of the transition into the next developmental phase of the methodology are already being assessed. From the pioneering work describing the regeneration of solely albino plantlets or very erratic genotype-dependent regeneration of a small number of haploid green plants, anther culture methodology is now moving into the phase of generating many haploids for certain genotypes. However, the application of anther culture methodology in ryegrass and fescue breeding will only be successful if the technique allows for recovery of sufficient numbers of doubled-haploid green plants from a broad spectrum of commercially-useful germplasm. Along these lines, the knowledge gained on the genetic control of the androgenetic response in perennial ryegrass opened up opportunities for evaluation of strategies based on the incorporation of inducer genotypes in ongoing breeding programs. First published results indicate improved rates of green plant production in hybrids with breeding lines. The requirement for an increased responsiveness to anther culture in agronomically acceptable germplasm seems in principle feasible. If fertile, highly inbred lines from doubled haploids can be established in *L. perenne*, it seems likely that synthetic varieties or F_1 hybrids based on these doubled haploids may be released in the not too distant future.

References

Abdullah AA, Pedersen S, Andersen SB (1994) Triploid and hexaploid regenerants from hexaploid timothy (*Phleum pratensis* L.) via anther culture. Plant Breed 112: 342–345

Ban Y, Kokubu T, Miyaji Y (1971) Production of haploid plant by anther culture of *Setaria italica*. Bull Fac Agri Kagoshima Univ 21 :77–80

Bante I, Sonke T, Tandler RF, van den Bruel AMR, Meyer EM (1990) Anther culture of *Lolium perenne* and *Lolium multiflorum*. In: Sangwan RS, Sangwan-Norreel BS (eds) The impact of biotecnology in agriculture. Kluwer Academic, Dordrecht, pp 105–127

Blaydes DF (1966) Interaction of kinetin and various inhibitors in the growth of soybean tissues. Physiol Plant 19: 748–753

Boppenmeier J, Züchner S, Foroughi-Wehr B (1989) Haploid production from barley yellow dwarf virus resistant clones of *Lolium*. Plant Breed 103: 216–220

Buckner RC, Hill HD, Burrus PB (1961) Some characteristics of perennial and annual ryegrass × tall fescue hybrids and of the amphidiploid progenies of annual ryegrass × tall fescue. Crop Sci 1: 75–80

Bui Dang Ha D, Pernes J (1982) Androgenesis in pearl millet I. Analysis of plants obtained from microspore culture. Z Pflanzenphysiol 108: 317–327

Clapham D (1971) In vitro development of callus from the pollen of *Lolium* and *Hordeum*. Z Pflanzenzucht 65: 285–292

Creemers-Molenaar J, Beerepoot LJ (1992) In vitro culture and micropropagation of ryegrass (*Lolium* spp.). In: Bajaj YPS (ed) High-Tech and micropropagation III. Biotechnology inagriculture and forestry, vol 19. Springer, Berlin Heidelberg New York, pp 549–575

De Buyser J, Henry Y, Lonnet P, Hertzog R, Hespel A (1987) "Florin": a doubled haploid wheat variety developed by the anther culture method. Plant Breed 98: 53–56

Eizenga GC, Kasperbauer MJ (1985) Chromosome pairing in tall fescue haploids derived by anther-panicle culture. J Hered 76: 99–102

Foroughi-Wehr B, Wenzel G (1993) Andro- and parthenogenesis. In: Hayward MD, Bosemark NO, Romagosa I (eds) Plant breeding, principles and prospects. Chapman and Hall, London, pp 261–277

Foroughi-Wehr B, Mix G, Gaul H, Wilson HM (1976) Plant production from cultured anthers of *Hordeum vulgare* L. Z Pflanzenzucht 77: 198–204

Genovesi AD (1990) Maize (*Zea mays* L.): in vitro production of haploids. In: Bajaj YPS (ed) Haploids in crop improvement I. Biotechnology in agriculture and forestry, vol 12. Springer, Berlin Heidelberg New York, pp 176–203

Guha S, Maheshwari SC (1964) In vitro production of embryos from anthers of *Datura*. Nature 204: 497

Halberg N, Olesen A, Tuvesson IKD, Andersen SB (1990) Genotypes of perennial ryegrass (*Lolium perenne* L.) with high anther culture response through hybridization. Plant Breed 105: 89–94

Haydu Z, Vasil IK (1981) Somatic embryogenesis and plant regeneration from leaf tissues and anthers of *Pennisetum purpureum* Schum. Theor Appl Genet 59: 269–273

Hayward MD, Olesen A, Due IK, Jenkins R, Morris P (1990) Segregation of isozyme marker loci amongst androgenetic plants of *Lolium perenne* L. Plant Breed 104: 68–71

Huang CS, Tsay HS, Chern CG, Chen CC, Yeh CC, Tseng TH (1988) Japonica rice breeding using anther culture. J Agric Res China 37: 1–8

Jauhar PP (1993) Cytogenetics of the *Festuca-Lolium* complex. Relevance to breeding. In: Frankel R, Grossman M, Linskens HF, Maliga P, Riley R (eds) Monographs on theoretical and applied genetics, vol 18. Springer, Berlin Heidelberg New York, 243 pp

Kasperbauer MJ (1990a) Haploids: derivation and evaluation. In: Kasperbauer MJ (ed) Biotechnology in tall fescue improvement. CRC Press, Boca Raton, pp 79–96

Kasperbauer MJ (1990b) Doubled haploids: derivation and evaluation. In: Kasperbauer MJ (ed) Biotechnology in tall fescue improvement. CRC Press, Boca Raton, pp 97–108

Kasperbauer MJ, Eizenga GC (1985) Tall fescue doubled haploids via tissue culture and plant regeneration. Crop Sci 25: 1091–1095

Kasperbauer MJ, Buckner RC, Springer WD (1980) Haploid plants by anther-panicle culture of tall fescue. Crop Sci 20: 103–107

Kasperbauer MJ, Karlen DL, Burton HR (1987) Ploidy effects on protein, in vitro dry matter disappearance, and potassium/(calcium + magnesium) equivalent ratio in tall fescue forage. Crop Sci 27: 1081–1082

Kaul K (1990) Potential biotechnological approaches. In: Kasperbauer MJ (ed) Biotechnology in tall fescue improvement. CRC Press, Boca Raton, pp 13–23

Linsmaier EM, Skoog F (1965) Organic growth factor requirements of tobacco tissue culture. Physiol Plant 18: 100–127

Madsen S, Olesen A, Andersen SB (1993) Self-fertile doubled haploid plants of perennial ryegrass (*Lolium perenne* L.). Plant Breed 110: 323–327

Madsen S, Olesen A, Dennis B, Andersen SB (1995) Inheritance of anther culture response in perennial ryegrass (*Lolium perenne* L.). Plant Breed 114: 165–168

Marburger JE, Wang RRC (1988) Anther culture of some perennial triticeae. Plant Cell Rep 7: 313–317

Miller CO (1963) Kinetin and kinetin-like compounds. In: Linskens HF, Tracey MV (eds) Moderne Methoden der Pflanzenanalyse, vol 6. Springer, Berlin Heidelberg New York, pp 192–202
Morrison RA, Evans DA (1988) Haploid plants from tissue culture: new plant varieties in shortened time frame. Bio/Technology 6: 684–690
Niizeki M (1977) Haploid, polyploid and aneuploid plants from cultured anthers and calluses in species of *Nicotiana* and forage crops. J Facul Agri Hokkaido Univ 58: 343–466
Niizeki M, Kita F (1973) Studies on plant cell and tissue culture. III. In vitro induction of callus from anther culture of forage crops. J Facul Agri Hokkaido Univ 57: 293–300
Nitsch C (1977) Culture of isolated microspores. In: Reinert J, Bajaj YPS (eds) Applied and fundamental aspects of plant cell, tissue, and organ culture. Springer, Berlin Heidelberg New York, pp 268–278
Nitsch JP, Nitsch C (1969) Haploid plants from pollen grains. Science 163: 85–87
Nitzsche W (1970) Herstellung haploider Pflanzen aus *Festuca-Lolium* Bastarden. Naturwissenschaften 57: 199–200
Nitzsche W, Wenzel G (1977) Haploids in plant breeding. Parey, Berlin, pp 46–49
Olesen A, Andersen SB, Due IK (1988) Anther culture response in perennial ryegrass (*Lolium perenne* L.). Plant Breed 101: 60–65
Opsahl-Ferstad HG, Bjørnstad A, Rognli OA (1994a) Influence of medium and cold pretreatment on androgenetic response in *Lolium perenne* L. Plant Cell Rep 13: 594–600
Opsahl-Ferstad HG, Bjørnstad A, Rognli OA (1994b) Genetic control of androgenetic response in *Lolium perenne* L. Theor Appl Genet 89: 133–138
Pagniez M, Demarly Y (1979) Obtention d'individus androgénétiques par culture in vitro d'anthères de ray-grass d'Italie (*Lolium multiflorum* Lam.). Ann Amélior Plantes 29: 631–637
Picard E, Rode A, Doussinault G, Rousset M, Rives M (1990) Wheat (*Triticum aestivum*): in vitro production and utilization of doubled haploids. In: Bajaj YPS (ed) Haploids in crop improvement I. Biotechnology in agriculture and forestry, vol 12. Springer, Berlin Heidelberg New York, pp 101–124
Reinert J, Bajaj YPS (1977) Anther culture: haploid production and its significance. In: Reinert J, Bajaj YPS (eds) Applied and fundamental aspects of plant cell, tissue, and organ culture. Springer, Berlin Heidelberg New York, pp 251–267
Rose JB, Dunwell JM, Sunderland N (1987a) Anther culture of *Lolium temulentum*, *Festuca pratensis* and *Lolium* × *Festuca* hybrids. I. Influence of pretreatment, culture medium and culture incubation conditions on callus production and differentiation. Ann Bot 60: 191–201
Rose JB, Dunwell JM, Sunderland N (1987b) Anther culture of *Lolium temulentum*, *Festuca pratensis* and *Lolium* × *Festuca* hybrids. II. Anther and pollen development in vivo and in vitro. Ann Bot 60: 203–214
Saito K, Nakayama R, Takeda K, Kuwata H (1973) Studies on the breeding of the grass IV. Differentiation of plants by anther culture in orchardgrass and smooth bromegrass. Bull Fac Agri Hirosaki Univ 21: 1–8
Stanis VA, Butenko RG (1984) Developing viable haploid plants in anther culture of ryegrass. Dokl Biol Sci 275: 249–251
Utz HF, Oettler G (1978) Performance of inbred lines and their top crosses in perennial ryegrass (*Lolium perenne* L.). Z Pflanzenzucht 80: 223–229
Wang XZ, Hu H (1984) The effect of potato II medium for triticale anther culture. Plant Sci Lett 36: 237–239
Wenzel G, Foroughi-Wehr B (1984) Anther culture of cereals and grasses. In: Vasil IK (ed) Cell culture and somatic cell genetics of plants, vol 1. Academic Press, Orlando, pp 311–327
Wenzel G, Frei U, Jahoor A, Graner A, Foroughi-Wehr B (1995) Haploids – an integral part of applied and basic research. In: Terzi M, Cella R, Falavigna A (eds) Current issues in plant molecular and cellular biology. Kluwer Academic, Dordrecht, pp 127–135
Xu ZH (1990) Barley (*Hordeum vulgare* L.): anther culture and production of haploids. In: Bajaj (ed) Haploids in crop improvement I. Biotechnology in agriculture and forestry, vol 12. Springer, Berlin Heidelberg New York, pp 125–175

Zenkteler M, Misiura E (1974) Induction of androgenic embryos from cultured anthers of *Hordeum*, *Secale* and *Festuca*. Biochem Physiol Pfl 165: 337–340

Zhu DY, Pan XG (1990) Rice (*Oryza sativa* L.): guan 18 – an improved variety through anther culture. In: Bajaj (ed) Haploids in crop improvement I. Biotechnology in agriculture and forestry, vol 12. Springer, Berlin Heidelberg New York, pp 204–211

Chapter 5
Plant Regeneration from Suspension Cells and Protoplasts

5.1 Introduction

Embryogenic callus cultures (Chap. 3) have proved invaluable for the development of regenerable cell suspension cultures. When callus is grown in an agitated liquid medium, it forms a suspension composed of cellular aggregates and free cells. Embryogenic suspension cultures of grasses are comprised mostly of pro-embryogenic cell clusters and groups of small, meristematic cells rich in cytoplasm and containing starch. Plating of suspension cells on solidified media is required for the formation of somatic embryos and plant regeneration (Vasil 1987).

Well established cell suspension cultures are still rather difficult to obtain and maintain. Critical to the success in establishing embryogenic suspension cultures is the early identification and selective enrichment of the embryogenic structures formed (Spangenberg 1995). It is indeed this critical issue which makes the establishment of embryogenic cell cultures an art that also depends upon parameters beyond experimental control (Potrykus 1990). Other parameters, such as explant source, genotype and culture conditions, play an important role in the establishment and regeneration of embryogenic cultures (Morrish et al. 1987; Vasil 1988).

The routine maintenance of established embryogenic suspension cultures is time-consuming and expensive, and there is a risk of loss of valuable materials through contamination, genetic changes etc. In addition, embryogenic cultures may lose their embryogenic regeneration potential with time. Cryopreservation has become a tool for long-term preservation of cell cultures, probably without genetic alterations (Kartha 1987). This technique of deep freezing, in which cells and tissues are frozen under controlled conditions and stored in liquid nitrogen at −196 °C, offers the advantage that both metabolic processes and biological deterioration are slowed considerably.

Protoplasts are cells deprived of their walls. Enzymatically isolated protoplasts represent an ideal system to demonstrate totipotency of single plant cells, since plant regeneration from cultured protoplasts is possible. In contrast to most dicot species, where protoplasts isolated from differentiated plant tissues have the ability to de-differentiate and re-enter the cell cycle, corresponding protoplasts from grasses lack this ability. In contrast, embryogenic

Table 5.1. Suspension and protoplast culture in forage and turf grasses

Species	Source	Culture response	References
Festuca arundinacea (tall fescue)	Protoplasts	Plants	Dalton (1988a, b)
	Protoplasts	Plants	Takamizo et al. (1990)
	Suspension	Plants	Rajoelina et al. (1990)
	Suspension and protoplasts	Plants	Wang et al. (1994)
	Suspension	Plants	Fournier et al. (1996)
Festuca pratensis (meadow fescue)	Suspension and protoplasts	Fertile plants	Wang et al. (1993a)
Festuca rubra (red fescue)	Suspension	Plants	Zaghmout and Torello (1989, 1992a)
	Protoplasts	Morphogenic calli	Zaghmout and Torello (1990a)
	Suspension and protoplasts	Plants	Spangenberg et al. (1994)
Lolium multiflorum (Italian ryegrass)	Protoplasts	Calli	Jones and Dale (1982)
	Protoplasts	Plants	Dalton (1988b)
	Suspension	Plants	Rajoelina et al. (1990)
	Suspension and Protoplasts	Fertile plants	Wang et al. (1993b, 1995)
Lolium perenne (perennial ryegrass)	Protoplasts	Plantlets	Dalton (1988a, b)
	Suspension and protoplasts	Plants	Creemers-Molenaar et al. (1989)
	Suspension	Plantlets	Zaghmout and Torello (1990b)
	Protoplasts	Calli	Creemers-Molenaar et al. (1992)
	Protoplasts	Albino plantlets	Zaghmout and Torello (1992b)
	Suspension and protoplasts	Fertile plants	Wang et al. (1993b, 1995)
	Suspension	Fertile plants	Altepeter and Posselt (1994)
	Suspension and protoplasts	Plants	Olesen et al. (1995)
Lolium × *boucheanum* (Hybrid ryegrass)	Suspension and protoplasts	Plants	Wang et al. (1993b)
Lolium temulentum	Suspension	Morphogenic calli	Dalton and Thomas (1992)
Agrostis alba (redtop)	Suspension and protoplasts	Plants	Asano and Sugiura (1990)

Agrostis palustris (creeping bentgrass)	Suspension and protoplasts	Plants	Terakawa et al. (1992)
Bromus inermis (bromegrass)	Suspension	Albino plantlets	Gamborg et al. (1970)
	Protoplasts	Plantlets	Kao et al. (1973)
Dactylis glomerata (cocksfoot)	Suspension	Plants	Gray et al. (1984)
	Suspension and protoplasts	Plants	Horn et al. (1988)
Elymus giganteus (wildrye)	Suspension	Plants	Wang et al. (1996)
Panicum maximum (guinea grass)	Protoplasts	Plantlets	Lu et al. (1981)
Panicum miliaceum (proso millet)	Protoplasts	Albino plantlets	Heyser (1984)
Paspalum dilatatum (dallisgrass)	Suspension and protoplasts	Plants	Akashi and Adachi (1992)
Paspalum notatum (bahiagrass)	Suspension	Plants	Akashi et al. (1993)
Paspalum scrobiculatum (minor millet)	Suspension	Plants	Nayak and Sen (1989)
Pennisetum americanum (pearl millet)	Protoplasts	Plantlets	Vasil and Vasil (1980)
Pennisetum purpureum (napier grass)	Protoplasts	Plantlets	Vasil et al. (1983)
	Suspension and protoplasts	Plants	Wan and Vasil (1996)
Poa pratensis (Kentucky bluegrass)	Protoplasts	Albino plantlets	Van der Valk et al. (1988)
	Suspension	Plants	Nielsen and Knudsen (1993)
	Protoplasts	Plants	Nielsen et al. (1993)
Zoysia japonica (Japanese lawngrass)	Protoplasts	Plants	Inokuma et al. (1996)

cell suspensions have proven a unique source of totipotent protoplasts in grasses (Potrykus and Shillito 1986; Vasil 1987, 1988; Roest and Gilissen 1989).

Protoplasts can be subjected to fusion and direct gene transfer treatments for the production of somatic hybrids (Chap. 6) and transgenic plants (Chap. 7), respectively. In addition, suspension cultures can be used directly as targets for genetic transformation experiments to recover transgenic grass plants based on protoplast-independent gene transfer methods such as biolistic and silicon carbide whisker-mediated transformation (Chap. 8). The genetic manipulation at the cellular level of forage and turf grasses thus requires efficient means for the regeneration of whole plants from embryogenic suspension cultures and suspension-derived protoplasts. Reports on establishment of and plant regeneration from embryogenic cell suspensions and corresponding protoplasts in forage and turf grasses are discussed below (Table 5.1).

5.2 Cell Suspension and Protoplast Cultures in *Festuca*

5.2.1 Plant Regeneration from Embryogenic Cell Suspensions in *Festuca*

The establishment of embryogenic suspension cultures allowing plant regeneration from plated suspension cells was first described for *F. arundinacea* (Dalton 1988a, b). In these studies, embryogenic cell suspensions were initiated from up to 30 mature zygotic embryos cultured in liquid Murashige and Skoog (MS) medium supplemented with 3% sucrose and 10 mg/l 2,4-dichlorophenoxyacetic acid (2,4-D). Since mixed populations of zygotic embryos from one cultivar were used as explants to initiate the cultures, multiple genotype-derived cell suspensions were obtained. These suspension cultured cells were initially capable of regenerating green shoots when plated on media for morphogenesis, but their regenerative ability quickly declined with age (Dalton 1988a, b). Detailed information on frequency of plant regeneration from suspension cells was not provided in these early reports.

Two protocols for establishing multiple genotype-derived embryogenic cell suspensions of *F. arundinacea* using short-term (5–8 weeks) cultured calli initiated either from immature embryos or from mature seeds were compared by Rajoelina et al. (1990). Plant regeneration frequencies between 50 and 73% were observed. Green and albino plantlets were regenerated from the plated suspension cells. The green:albino ratio was 9:1. The suspension cultures were only able to retain their embryogenic capacity for up to 13 weeks (Rajoelina et al. 1990).

A different approach using long-term (over 4 years) cultured, highly embryogenic calli has been reported for the initiation of embryogenic cell suspensions in *F. rubra* (Zaghmout and Torello 1989). The basal medium used for establishing these multiple genotype-derived suspension cultures consisted of half-strength MS salts supplemented with B5 vitamins, 3% sucrose, 3 g/l casein hydrolysate and 4 mg/l 2,4-D. The addition of casein hydrolysate to the medium was found to significantly enhance cell suspension growth rates. Plant regeneration was attained after transferring embryogenic cell clusters to hormone-free, half-strength MS medium. Red fescue plants regenerated from suspension cells survived the transfer to soil. No information on regeneration frequencies and green : albino ratios was given (Zaghmout and Torello 1989). However, a decline in regeneration ability and green : albino ratio was noted with time for these long-term cultures of red fescue in a later report (Zaghmout and Torello 1992a). A pretreatment with elevated levels of sucrose (up to 6.2%) was found to enhance regeneration capacity and to decrease the number of albino plantlets from both long-term callus and suspension cultures of red fescue (Zaghmout and Torello 1992a). Cell suspension growth was greatest when 4.6% sucrose was added to the pretreatment medium. High concentrations of sucrose in the hormone-free regeneration medium composed of half-strength MS salts and B5 vitamins were necessary to allow for efficient plantlet regeneration from the high-sucrose pretreated suspension cell aggregates (Zaghmout and Torello 1992a).

The use of single genotype-derived cell suspensions (Takamizo et al. 1990) opened up opportunities for the establishment of efficient plant regeneration systems applicable to a wide range of cultivars of different *Festuca* species (Wang et al. 1993a, 1995; Spangenberg et al. 1994). This approach was based on an initial screening of plated seeds for the induction of embryogenic callus to identify responsive genotypes. Individually subcultured single seed-derived embryogenic callus material was then used to initiate morphogenic suspension cultures from single genotypes in *F. pratensis*, *F. arundinacea* and *F. rubra*. Depending on the cultivar used, 3–20% of the seeds screened produced yellowish friable callus suitable for the initiation of single genotype-derived suspension cultures in all three *Festuca* species evaluated (Fig. 5.1; Wang et al. 1993a, 1995; Spangenberg et al. 1994).

Single genotype-derived suspension cultures were initiated and maintained in liquid AA medium (Müller and Grafe 1978) supplemented with 2% sucrose, 2% sorbitol and 1.5 mg/l 2,4-D. Embryogenic cell suspensions were established out of at least every second cultivar tested (Fig. 5.1). The embryogenic potential of the established cell suspensions was evaluated after 2- to 3-week proliferation on MS medium containing 3% sucrose and 2 mg/l 2,4-D prior to plating onto hormone-free MS regeneration medium. Frequencies of green plant regeneration (number of regenerable calli per number of plated calli from the best responding single genotype cell suspension within one cultivar) ranging between 30 and 80% were observed in *F. pratensis* (Fig. 5.1A; Wang et al. 1993a). Corresponding values for *F. arundinacea* were in the range of 50 to 80% (Fig. 5.1B). Both high (>90%) and low (<30%) regeneration

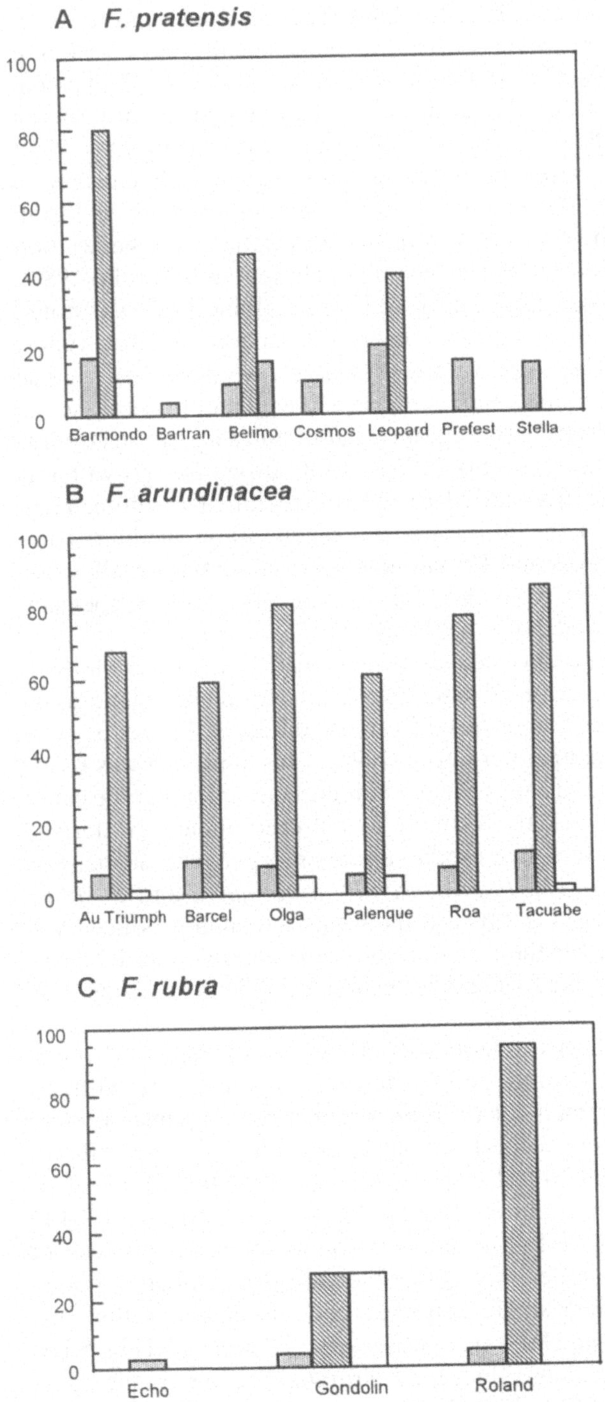

Fig. 5.1A–C. Initiation, establishment of, and plant regeneration from embryogenic suspension cultures in the fescues **A** *F. pratensis*, **B** *F. arundinacea* and **C** *F. rubra. Solid grey columns* indicate the number of embryogenic calli/number of plated seeds; *shaded columns* the number of green plantlets/ number of plated calli from cell suspensions; and *white columns* the number of albino plantlets/number of plated calli from cell suspensions. Frequencies are given as %

frequencies were obtained for individual cell suspensions of different cultivars in *F. rubra* (Fig. 5.1C; Spangenberg et al. 1994). Green:albino ratios varied, depending on the cultivar, between 1:0 and 1:1 (Fig. 5.1).

In the case of the major *Festuca* species, the established embryogenic suspension cultures retained their potential for regeneration of green plants when evaluated over a period of 8–14 months (Wang et al. 1993a; Spangenberg et al. 1994). Nevertheless, in order to ensure a constant long-term availability of highly regenerable defined suspension cultures, experimental conditions for the cryopreservation under liquid nitrogen of 4- to 6-month-old, single genotype-derived embryogenic cell suspensions were partially optimized (Wang et al. 1994). Different cryopreservation parameters, such as cryoprotectant composition, pre-freezing osmotic treatment of suspension cultures, time point of cryoprotectant addition, and post-thaw washing of cryopreserved cultures before plating, were evaluated. Significant differences in post-thaw growth of cryopreserved cell suspensions were detected between the different treatments. The use of 10% dimethyl sulfoxide (DMSO) and 0.5M sorbitol allowed for better recovery frequencies than the use of 5% DMSO, 10% glycerol and 8% sucrose. The preculture of embryogenic suspension cultures for 3–4 days in a culture medium containing 0.33M sorbitol (instead of the standard 0.11M) prior to cryopreservation improved post-thaw recovery of frozen cultures for most cryoprotectants tested (Fig. 5.2A–C). The time point of cryoprotectant addition played a role in post-thaw recovery of frozen embryogenic suspensions in the *Festuca* species considered (Fig. 5.2D, E). Washing of cryopreserved-thawed suspension cultures before plating was deleterious, particularly for *F. pratensis*, where a drastic drop in recovery frequencies was observed (Fig. 5.2F).

In general terms, for the cryogenic storage of embryogenic suspension cells, the use of 10% DMSO and 0.5M sorbitol as cryoprotectant, a 30-min cryoprotectant uptake time, and the use of a programmable freezing unit to guarantee low cooling rates (e.g. 1 °C/min) including a 1-h holding period at the transfer temperature of −40 °C were recommended (Wang et al. 1994).

Using this protocol, single genotype-derived embryogenic cell suspensions of *F. pratensis*, *F. arundinacea* and *F. rubra* were cryopreserved for 2–4 weeks, and their potential for plant regeneration was evaluated after post-thaw proliferation. No reduction in plant regeneration frequencies and green:albino ratios from these short-term cryopreserved and thawed cultures was found when compared with the respective non-frozen original cultures. Analogous experiments comparing single genotype-derived suspension cells cryopreserved for 6 months with corresponding maintained cultures revealed a relatively low (28%) regeneration frequency of the latter in contrast to a high (82%) in vitro regeneration capacity of the former. Adult plants regenerated from cryopreserved and maintained cultures were established under field conditions in order to analyze their agronomic performance (Fig 5.2G–J; Wang et al. 1994).

5.2.2 Plant Regeneration from Protoplasts in *Festuca*

Protoplast culture leading to plant regeneration in *F. arundinacea* was first reported by Dalton (1988a, b). Embryogenic cell suspensions initiated from a mixture of genotypes of one cultivar were used as the source of totipotent protoplasts in this early study. Protoplast yield was low for newly initiated suspension cultures, and reached 3×10^6 protoplasts/g fresh weight cells after

20 weeks of subculture. Colonies were formed when protoplasts isolated from morphogenic cell supensions were cultured in MS medium supplemented with 0.38 M glucose, 0.27 M mannitol, 0.2 mg/l 2,4-D and 0.1 mg/l benzylamino purine (BAP). Plating efficiencies of 0.1–0.4% were achieved. Up to 10% of the protoplast-derived colonies developed multiple shoots, and the green: albino ratio was 1:2. The in vitro morphogenic response of the protoplast-derived colonies was the same as for colonies derived from the respective cell suspensions. Soil-grown tall fescue plants from protoplasts were established. No information on fertility of the protoplast-derived plants was provided (Dalton 1988a,b).

Takamizo et al. (1990) isolated protoplasts from single genotype-derived suspension cultures of two cultivars in *F. arundinacea*. The yield of protoplasts varied from 10^6 to 10^7 protoplasts/g fresh weight cells. The AA medium (Müller and Grafe 1978) supplemented with 0.6 M glucose and 1 mg/l 2,4-D was successfully used for protoplast culture. Protoplast-derived colonies were obtained only when protoplasts were cultured in a bead type system including nurse cells. Plating efficiencies were in the range of 0.03–0.55%. Protoplast-derived colonies regenerated green shoots in 9.6% of the cases, and the green: albino ratio was 1:3. Here again, adult protoplast-derived plants of tall fescue were established in soil (Takamizo et al. 1990).

The protocol developed by Takamizo et al. (1990) allowed the establishment of an improved protoplast culture system applicable to a wide range of cultivars in *F. arundinacea* (Fig. 5.1B; Wang et al. 1995). Protoplast plating efficiencies between 0.01 and 0.1% were achieved. Relatively high plant regeneration frequencies in the range of 40–70% were observed, and green: albino ratios varied between 1:0 and 6:1. The protoplast-derived plants survived transfer to soil and were established under greenhouse and field conditions. Similar behaviour (regeneration frequencies and green: albino ratios) and cultivar dependence were apparent when comparing plant regeneration

Fig. 5.2A–J. Cryopreservation of embryogenic suspension cultures in fescues. **A** Growth of control single genotype-derived cell suspension of *F. arundinacea* after plating onto proliferation medium. **B** Post-thaw growth of cell suspension of *F. arundinacea* (same genotype as in **A** precultured in AA medium containing 0.33 M sorbitol and cryopreserved using 10% DMSO and 0.5 M sorbitol as cryoprotectant. **C** Post-thaw growth of cell suspension of *F. arundinacea* (same genotype as in **A** precultured in AA medium containing 0.11 M sorbitol and cryopreserved using 5% DMSO, 10% glycerol and 8% sucrose as cryoprotectant. **D** Post-thaw growth of single genotype-derived cell suspension of *F. pratensis* cryopreserved 30 min after cryoprotectant (10% DMSO and 0.5 M sorbitol) addition. **E** Post-thaw growth of cell suspension of *F. pratensis* (same genotype as in **D** cryopreserved immediately after addition of cryoprotectant (10% DMSO and 0.5 M sorbitol). **F** Post-thaw growth of cell suspension of *F. pratensis* (same genotype as in **D**) after washing off the added cryoprotectant. **G** Regeneration from control (not cryopreserved) embryogenic cultures of *F. arundinacea*. **H** Regeneration from cryopreserved-thawed embryogenic cultures of *F. arundinacea*. **I, J** Plants regenerated from cryopreserved-thawed embryogenic cultures of *F. arundinacea* growing under greenhouse (**I**) and field (**J**) conditions. (Wang et al. 1994)

directly from embryogenic suspension cultures with regeneration from the corresponding protoplasts (Wang et al. 1995).

The culture of protoplasts from multiple genotype-derived suspension cells of *F. rubra* has been described by Zaghmout and Torello (1990a). Different protoplast isolation and culture conditions were tested to partially optimize plating efficiencies. Protoplast-derived colonies were obtained only when nurse cells were used. The addition of 2% DMSO to the protoplast culture medium led to a fourfold increase in plating efficiencies which reached values of up to 1%. However, no protoplast-derived red fescue plant was obtained in this study (Zaghmout and Torello 1990a).

By the use of highly morphogenic single genotype-derived cell suspensions, the recovery of a large number of *F. rubra* plants from protoplasts was possible (Spangenberg et al. 1994). Protoplasts formed microcalli when cultured in agarose beads and when nurse cells were initially added to the surrounding liquid medium in bead-type culture. Plating efficiences were in the range of 0.1–0.3%. More than 85% of the protoplast-derived microcalli regenerated plants that were established in soil and grown until maturity (Spangenberg et al. 1994). The performance of red fescue plants regenerated from suspension cells and corresponding protoplasts was evaluated in a 2-year field experiment (Stadelmann et al. 1997). Most plants showed a delay in inflorescence emergence, a reduced seed yield and differences in morphological traits when compared with seed-grown plants. However, regenerants showing similar or even superior performance to the seed-grown plants were also found (Stadelmann et al. 1997).

The generation of fertile plants from protoplasts of *F. pratensis* was reported by Wang et al. (1993a). Pure preparations of protoplasts (Fig. 5.3C) nearly free of contaminating undigested cells were routinely obtained with a yield of $0.2–1 \times 10^6$ protoplasts/g fresh weight cells. The overall plating efficiency was in the range of 0.01–0.1%, provided nurse cells were used during the first week in bead-type culture. Colonies growing in the agarose beads were visible after 3 weeks of culture and generated a lawn on the agarose-solidified medium after 4 weeks (Fig. 5.3D). Plant regeneration frequencies

Fig. 5.3A–L. Plant regeneration from embryogenic suspension-derived protoplasts in *Festuca pratensis* Huds. **A** Single genotype-derived embryogenic cell suspension of *F. pratensis* cv. Barmondo used for protoplast isolation 5 months after initiation of the culture. **B** In vitro regeneration of green and albino plantlets from morphogenic suspension culture of **A**. **C** Newly isolated protoplasts from morphogenic cell suspension. **D** Bead-type culture of meadow fescue protoplasts forming visible colonies 4 weeks after embedding. **E** In vitro plantlet regeneration from protoplast-derived callus 3 months after protoplast isolation. **F** Rooted plantlets regenerated from embryogenic cell suspension-derived protoplasts. **G** Greenhouse-growing vernalized plants regenerated from protoplasts. **H** Inflorescence of protoplast-derived meadow fescue plant with protruding anthers. **I** Lugol-stained pollen collected from anthers of protoplast-derived plant. **J** Mature plants regenerated from protoplasts setting seeds after crossing. **K** Germinating seeds collected from protoplast-derived plant. **L** Metaphase in meristem cell prepared from one representative meadow fescue plant regenerated from protoplasts. (Wang et al. 1993a)

A
B
C
D
E
F
G
H
I
J
K
L

from protoplast-derived calli were in the range of 20–30%. Only a few albino plantlets were regenerated, with green : albino ratios varying between 1:0 and 2:1. Protoplast-derived plants of meadow fescue were established in the greenhouse and were brought to flower. They developed normal inflorescences with protruding anthers containing up to 40% viable pollen (Fig. 5.3H, I). Crossings performed with these cytologically normal (2n = 14) protoplast-derived plants led to the production of viable seeds (Fig. 5.3J–L; Wang et al. 1993a). Furthermore, the genetic stability of the cell suspension- and protoplast-derived plants of meadow fescue regenerated in this study was assessed by RFLP and RAPD analyses (Vallés et al. 1993; Wang et al. 1993a). No evidence supporting the occurrence of somaclonal variation was detected (Chap. 9).

The in vitro performance of protoplasts isolated from long-term cryopreserved, thawed and re-established embryogenic cell suspensions of different *Festuca* species has been studied (Spangenberg et al. 1994; Wang et al. 1994). In *F. arundinacea*, protoplast yield and protoplast plating efficiency of suspension cultures re-established after 6-month cryostorage in liquid nitrogen were comparable to the original values for maintained cultures of equivalent age. A relatively high (76%) plant regeneration frequency was obtained for protoplasts isolated from the 6-month cryopreserved and re-established suspension culture, while a reduction in the capacity to regenerate green plantlets in vitro (33% regeneration frequency) from protoplasts isolated from the maintained suspension was observed. Plantlets regenerated from protoplasts isolated from cryopreserved suspension cultures were transferred to soil and grew normally under greenhouse conditions (Wang et al. 1994).

5.3 Cell Suspension and Protoplast Cultures in *Lolium*

5.3.1 Plant Regeneration from Embryogenic Cell Suspensions in *Lolium*

Plant regeneration from embryogenic cell suspensions in ryegrasses was first achieved for *L. perenne* (Dalton 1988a, b). In this early report the embryogenic suspension cultures were established from mixtures of genotypes of one perennial ryegrass cultivar. The same culture media that proved successful for tall fescue were used. Although no quantitative data were given on overall plant regeneration frequencies, the number of shoots produced from the regenerable cell suspensions was found to decline with increasing age of the cultures. These multiple genotype-derived cell suspensions regenerated both green and albino plantlets. The proportion of regenerated albino plantlets increased with age of the suspension culture. The green : albino ratio varied from 1:0.9 at the age of 13 weeks to 1:8 when the cell suspensions were 24 weeks old (Dalton 1988a, b).

Following the protocol developed by Dalton (1988a, b), plant regeneration from multiple genotype-derived suspension cultures initiated from mature embryos was later reported for different cultivars of *L. perenne* (Creemers-Molenaar et al. 1989; Altpeter and Posselt 1994). Alternatively, compact calli obtained from immature inflorescences were used to initiate embryogenic cell suspensions in perennial ryegrass (Creemers-Molenaar et al. 1989). Finely dispersed embryogenic suspension cultures were obtained from five out of the six cultivars tested. Up to 200 green shoots/g fresh weight of plated suspension cells were initially recovered, but the capacity to regenerate green shoots also decreased with time. In the case of most cell suspensions the regeneration potential was totally lost on average 25 weeks after initiation of the cultures (Creemers-Molenaar et al. 1989). Similar regeneration frequencies (206 green shoots/g fresh weight cells) of newly initiated fine cell suspensions have been reported for another perennial ryegrass cultivar (Altpeter and Posselt 1994). Mainly green plantlets and a few albinos were recovered in this study. Some of the green plantlets were transferred to soil, grown until flowering and seed was set (Altpeter and Posselt 1994).

Zaghmout and Torello (1990b) reported the establishment of multiple genotype-derived suspension cultures using embryogenic calli initiated from mature caryopses of one cultivar in *L. perenne*. In order to obtain highly regenerable suspension cultures, embryogenic cell types were selected either by discontinuous Percoll gradients or by filtering the newly initiated cultures. The addition of various components to the proliferation and regeneration media, such as proline, casein hydrolysate, coconut water, cefotaxime, fluridone and silver nitrate, was tested. Although regenerable suspension cultures were obtained, no significant beneficial effect on regeneration of green plants by the addition of these media constituents was observed. Most regenerated plantlets were albino, and only a few in vitro-growing green plantlets were finally recovered (Zaghmout and Torello 1990b).

Analogous experiments assessing the effects of various factors on proliferation and morphogenesis from cell suspension cultures have been reported for *L. temulentum* (Dalton and Thomas 1992). The treatments also included the addition of proline, cefotaxime, silver nitrate and casein hydrolysate, and further different gelling agents and carbohydrates were considered. No treatment significantly improved the proportion of colonies with globular or mature embryoids. However, the use of maltose-based medium increased frequencies of shoot differentiation from plated suspension cells from 47 to 69%. Albino shoots regenerated in 0.1% of the cases and the addition of proline to the medium increased this value to 0.8%. Only morphogenic calli differentiating shoots or roots were obtained from the plated suspension cells, but no adult soil-grown plant was recovered (Dalton and Thomas 1992).

A rapid loss of morphogenic capacity and sporadic regeneration is thus a reproducible feature in most studies on plant regeneration from multiple genotype-derived suspension cultures in ryegrasses (Dalton 1988a, b; Creemers-Molenaar et al. 1989; Zaghmout and Torello 1990b).

Single genotype-derived embryogenic suspension cultures have been the basis of a reproducible plant regeneration system established for different *Lolium* species and cultivars (Wang et al. 1993b). In this system, genotype

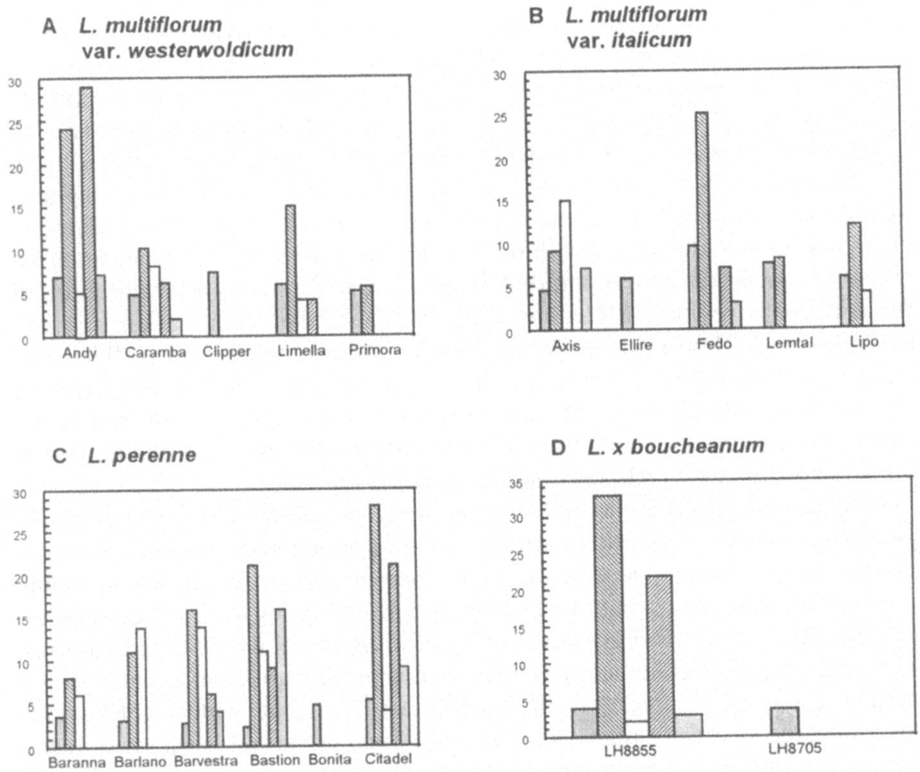

Fig. 5.4A–D. Initiation, establishment of, and plant regeneration from embryogenic cell suspensions and protoplasts in the ryegrasses **A** *L. multiflorum* var. *westerwoldicum*, **B** *L. multiflorum* var. *italicum*, **C** *L. perenne* and **D** *L.* x *boucheanum*. *Solid grey columns* indicate the number of embryogenic calli/number of plated seeds; *left-shaded columns* the number of green plantlets/number of plated calli from cell suspensions; *white columns* the number of albino plantlets/number of plated calli from cell suspensions; *right-shaded columns* the number of green plantlets/number of plated calli from protoplasts; and *dotted columns* the number of albino plantlets/number of plated calli from protoplasts. Frequencies are given as %. (Wang et al. 1993b)

screening for the induction of embryogenic calli was performed for *L. multiflorum* (Fig 5.4A, B), *L. perenne* (Fig. 5.4C) and *L.* × *boucheanum* (Fig. 5.4D). Single genotype-derived friable, yellowish callus was initiated in 2–8% of the plated seeds (Figs. 5.4, 5.5A) and corresponding embryogenic cell suspensions were established after 4–6 months (Fig. 5.5B). Plant regeneration from established embryogenic cell suspensions was achieved for three cultivars each in *L. multiflorum* var. *westerwoldicum* (Fig. 5.4A) and *L. multiflorum* var. *italicum* (Fig. 5.4B). Depending on the cultivar, green plantlets were regenerated in 5–25% of the plated suspension-derived calli (Fig. 5.4A, B), and corresponding green : albino ratios varied between 1 : 0 and 1 : 1.5. Multiple genotype-derived highly regenerable suspension cultures have also been obtained for one cultivar in *L. multiflorum* (Rajoelina et al. 1990).

Frequencies of regeneration of green plants varied between 53 and 58%. However, the embryogenic capacity of these suspension cultures could only be maintained for 9–13 weeks.

In *L. perenne*, single genotype-derived regenerable cell suspensions have been established in five out of six cultivars tested (Fig. 5.4C; Wang et al. 1993b). The frequency of green plant regeneration varied depending on the cultivar between 8 and 28% (Fig. 5.4C). Corresponding green : albino ratios were in the range of 0.8:1 to 6:1. In *L.* × *boucheanum*, embryogenic suspension cultures were established for single genotypes in one of the two breeding lines included in the genotype screening. More than 30% of the calli plated from this cell suspension regenerated green plants, and the green : albino ratio was 10:1 (Figs. 5.4D, 5.5D; Wang et al. 1993b).

A clear genotype dependence for obtaining appropriate callus to initiate regenerable suspension cultures has been observed in ryegrasses (Wang et al. 1993b). A detailed evaluation of genotype effects on plant regeneration from callus and cell suspension cultures has been carried out by Olesen et al. (1995) for *L. perenne*. Meristem-derived callus from individual clonally propagated plants was used to initiate cell suspensions. A total of 156 suspension cultures was established from 18 different perennial ryegrass clones. On average, plant regeneration frequency of the plated cell colonies was 7%, and the green : albino ratio was 1:1.3. Genotypes had a large impact on regeneration frequency and percentage of green plants from callus culture, where this component explained 59 and 83% of the total variation, respectively. However, since there was a large difference in the regeneration behaviour between suspensions originating from the same clone, the effect of genotype on regeneration from suspension cultures was smaller than the one oberserved for callus cultures. It accounted for 15% of the total variation observed for the suspension cultures. Genotypes showing high regeneration potential in callus culture also performed better in suspension culture (Olesen et al. 1995).

It has been a general phenomenon that frequency of plant regeneration as well as green : albino ratio declined over time when cell suspensions were maintained under normal conditions (Dalton 1988a, b; Creemers-Molenaar et al. 1989; Wang et al. 1993b). The decrease in the capacity for plant regeneration of suspension cultures in ryegrasses may in part be due to in vitro culture-induced chromosomal variation. Ploidy levels of suspension cells of *L. perenne* were found to change from euploidy to aneuploidy with increasing age of the suspension cultures (Creemers-Molenaar et al. 1992a). Furthermore, the frequency of differentiation of green shoots from plated suspension cells decreased simultaneously with the change in ploidy level. The majority of regenerated green shoots were diploid, whereas albino shoots obtained from the same suspension cultures were predominantly polyploid, mixploid and chimeric (Creemers-Molenaar et al. 1992a).

In order to maintain the morphogenic potential of suspension cultures of *L. perenne*, Creemers-Molenaar et al. (1992a) investigated the effects of cold storage on plant regeneration. Cold storage at 4 °C of embryogenic cell

Fig. 5.5A–J. Plant regeneration from embryogenic suspension-derived protoplasts in *Lolium perenne* L. **A** Single genotype-derived embryogenic callus of *L. perenne* cv. Citadel 2 months after plating seeds. **B** Embryogenic cell suspension culture of perennial ryegrass established from **A**. **C** Post-thaw growth of cryopreserved-thawed cell suspension 3 weeks after plating cells onto proliferation medium. **D** In vitro regeneration of green plantlets from cryopreserved-thawed suspension cells of **C**. **E** Newly isolated protoplasts from morphogenic cell suspension of **B**. **F** Bead-type culture of perennial ryegrass protoplasts forming visible colonies 3 weeks after embedding. **G** Regenerating protoplast-derived callus 3 months after protoplast isolation. **H** Rooted plantlets regenerated from embryogenic cell suspension-derived protoplasts and growing in vitro. **I** Plants of perennial ryegrass regenerated from protoplasts and growing in the greenhouse. **J** Protoplast-derived plants of perennial ryegrass growing in the field. (Wang et al. 1993b)

suspensions was useful for prolonging their regeneration capacity from 22 to 34 weeks. Suspension cultures maintained under standard conditions for 22 weeks had lost their competence for plant regeneration. Cold storage of suspension cultures at 4°C in the dark not only extended the period of their regeneration competence, but also increased regeneration frequency (Creemers-Molenaar et al. 1992a).

Protocols for long-term cryostorage under liquid nitrogen of regenerable cell suspensions which allow their efficient post-thaw recovery and plant regeneration after re-establishment of suspension cultures have been developed for *L. multiflorum*, *L. perenne* and *L.* × *boucheanum* (Wang et al. 1994). This study revealed that the preculture of ryegrass suspension cells in AA medium containing 0.33 M sorbitol prior to freezing and the use of 10% DMSO and 0.5 M sorbitol as cryoprotectant allowed 40–60% of the frozen and thawed cell aggregates to recover (Fig. 5.5C). In contrast to the results obtained for cryopreservation of suspension cells in *Festuca* (Sect. 5.3.1), the time point of cryoprotectant addition and post-thaw washing did not have a significant influence on recovery frequencies for cryopreserved suspension cells in *Lolium*. Plant regeneration frequencies and green:albino ratios of embryogenic cultures re-established after cryopreservation were similar to those of the original non-frozen controls (Wang et al. 1994).

5.3.2 Plant Regeneration from Protoplasts in *Lolium*

The culture of protoplasts isolated from cell suspensions in ryegrasses was first reported for *L. multiflorum* (Jones and Dale 1982). Although the conditions required for efficient protoplast division and colony formation were described in this early study, the callus lines derived from the cultured Italian ryegrass protoplasts did not lead to plant regeneration (Jones and Dale 1982).

Some improvements in establishing plant regeneration systems from Italian ryegrass protoplasts were made by Dalton (1988b) by using multiple genotype-derived embryogenic cell suspensions as a source of protoplasts. Plating efficiencies from the cultured protoplasts were less than 0.1%. Green shoots were regenerated with a low frequency from the protoplast-derived colonies obtained, and only five plants were established in soil.

An efficient system allowing fertile plant regeneration from protoplasts of *L. multiflorum* was first described by Wang et al. (1993b). The system was based on the use of single genotype-derived embryogenic cell suspensions as a source of totipotent protoplasts, cryopreservation for long-term availability of appropriate cultures and a protoplast bead-type culture system including nurse cells. Protoplast yield was in the range of 1–5 × 10^5 protoplasts/g fresh weight cells. An overall plating efficiency from 0.01 to 0.1% was obtained. Depending on the cultivar, 5–30% of the protoplast-derived calli regenerated green plantlets (Fig. 5.4A, B). Green:albino ratios ranged from 1:0 to 0:1. All protoplast-derived plants of Italian ryegrass transferred to soil flowered, were male and female fertile, set seeds, and 80% of the collected seeds could be germinated. In addition, these plants have been shown to be largely genetically stable at the loci screened by RAPD markers (Chap. 9).

The culture of cell suspension-derived protoplasts in *L. perenne* was first reported by Dalton (1988a,b). Protoplast yields of up to 3×10^6 protoplasts/g fresh weight cells were obtained. Plating efficiency of the cultured protoplasts was 1%. Although up to 10% of the transferred colonies developed multiple shoots, the majority (>99%) of the regenerants were albinos. Only two protoplast-derived colonies formed green shoots. However, no perennial ryegrass plant from protoplasts survived the transfer to soil (Dalton 1988a, b).

First soil-grown plants from suspension cell-derived protoplasts of *L. perenne* were recovered by Creemers-Molenaar et al. (1989). The protoplast culture response of multiple genotype-derived cell suspensions at different culture age was tested. Protoplast yield was 1×10^5 protoplasts/g fresh weight cells when the suspension cultures were 9–10 weeks old, while older (>20 weeks) cell suspensions yielded up to 1.4×10^7 protoplasts/g fresh weight cells. The fast-growing older suspension cultures yielded a large number of protoplasts which formed numerous microcalli in liquid and agarose-solidified media, but failed to regenerate plants. In contrast, protoplasts isolated from young suspension cultures of moderate growth rate were able to form a small number of regenerable microcalli. Out of 27 cell suspensions tested, one yielded regenerable protoplasts and finally led to the recovery of three soil grown protoplast-derived plants (Creemers-Molenaar et al. 1989).

In attempting to improve protoplast culture procedures in *L. perenne*, Creemers-Molenaar et al. (1992b) investigated the effects of conditioned medium, acetylsalicylic acid and antioxidants on the culture response of suspension-derived protoplasts. Conditioning of the protoplast culture medium was found to be indispensable for the proliferation of microcalli from protoplasts derived from young embryogenic suspension cultures. The addition to the protoplast isolation and culture medium of acetylsalicylic acid and antioxidants (superoxide dismutase, catalase, glutathione, glutathione peroxidase and phospholipase A2) improved plating efficiency, while vitamins C and E and n-propyl gallate had no effects on the culture response of perennial ryegrass protoplasts. No regeneration of protoplast-derived plants was reported (Creemers-Molenaar et al. 1992b).

Further attempts to improve the culture response of protoplasts in *L. perenne* were made by testing different protoplast plating techniques, including the use of liquid droplets, liquid layers, ultraclone growth chambers and mixed-nurse cultures (Zaghmout and Torello 1992b). Although protoplast divisions were observed in all culture methods tested, the formation of protoplast-derived colonies was achieved only when the latter two plating techniques were used. The highest plating efficiencies of up to 3% were achieved with the mixed-nurse plating technique. However, only albino plantlets were recovered in this study (Zaghmout and Torello 1992b).

The first fertile plants from protoplasts in *L. perenne* were obtained using an analogous culture method described for *L. multiflorum* by Wang et al. (1993b). Single genotype-derived cell suspension provided a suitable source of morphogenic protoplasts (Fig. 5.5B, E). Protoplast yields in the range of 1–5 × 10^5 protoplasts/g fresh weight suspension cells were routinely obtained. Plating efficiencies between 0.04 and 0.4% were achieved provided protoplasts were cultured in an agarose bead-type system including nurse cells during the first week of culture (Fig. 5.5F). A tenfold decrease in plating efficiency was observed when nurse cells were omitted. Depending on the cultivar, 6–22% of the protoplast-derived calli regenerated green plantlets (Figs. 5.4C, 5.5G, H). Green: albino ratios were in the range of 2:1 to 1:2. The protoplast-derived perennial ryegrass plants grown under greenhouse and field conditions showed normal phenotype and set seeds (Fig. 5.5I, J; Wang et al. 1993b, 1995).

The effects of the use of nurse cells and conditioned medium on colony formation and plant regeneration of *L. perenne* protoplasts have been investigated by Folling et al. (1995). Nurse cultures were found to improve the formation of macrocolonies from 0.17 to 1.7%. Self-conditioning, i.e. the use of conditioned medium taken from the protoplast-donor cell suspension, increased plating efficiency slightly. Conditioned medium harvested from other suspension cultures had on average a negative effect on colony formation and vitality. Nurse cultures and self-conditioning increased green: albino ratios from 1:1.4 to 18:1 (Folling et al. 1995).

Plant regeneration from protoplasts has also been reported for *L.* × *boucheanum* (Wang et al. 1993b). Plating efficiencies of hybrid ryegrass protoplasts of 0.04–0.4% were obtained when nurse cells were used. About 22% of the protoplast-derived calli regenerated green plantlets. Albino plantlets were also recovered, and the average green: albino ratio was 6:1 (Fig. 5.4D; Wang et al. 1993b).

The in vitro culture performance of protoplasts isolated from cryopreserved, thawed and re-established cell suspensions of *L. multiflorum* and *L. perenne* has been evaluated (Wang et al. 1993b, 1994). Protoplast yield, plating efficiency and green plant regeneration frequency were within the range observed for non-cryopreserved control suspension cultures of equivalent age prior to cryopreservation.

The fertility and growth in the field of *L. perenne* plants regenerated from suspension cultured cells and protoplasts, as well as the field performance of half-sib families from the suspension cell-derived plants were evaluated (Stadelmann et al. 1997, 1998). Phenology, morphology and fertility of regenerated *L. perenne* plants were studied for two growing seasons in a replicated field experiment. The performance of the plants regenerated from cryopreserved suspension cultures and protoplasts was similar to that of respective plants regenerated from routinely maintained suspension cultures. Although differences in morphology and phenology of the primary regenerants compared to seed-grown plants were observed, progenies of the regenerated plants did not differ from the respective controls. The potential of embryogenic suspension cultures and corresponding protoplasts for producing fertile, well-performing plants which can be integrated into breeding programs was demonstrated (Stadelmann et al. 1997, 1998).

5.4 Suspension and Protoplast Cultures in Other Grasses

Suspension and protoplast culture systems until the recovery of morphogenic calli and in vitro-grown plantlets have been described for forage and turf grasses other than fescues and ryegrasses, including *Bromus inermis* (Gamborg et al. 1970; Kao et al. 1973), *Elymus giganteus* (Wang et al. 1996), *Pennisetum americanum* (Vasil and Vasil 1980), *Pennisetum purpureum* (Vasil et al. 1983), *Panicum maximum* (Lu et al. 1981) and *Panicum miliaceum* (Heyser 1984; Table 5.1).

Regeneration of soil-grown plants from suspension cultures and corresponding protoplasts of species outside the *Festuca-Lolium* complex have been described for *Dactylis glomerata* (Gray et al. 1984; Horn et al. 1988), *Paspalum scrobiculatum* (Nayak and Sen 1989), *Agrostis alba* (Asano and Sugiura 1990), *Agrostis palustris* (Terakawa et al. 1992), *Paspalum dilatatum* (Akashi and Adachi 1992), *P. notatum* (Akashi et al. 1993), *Pennisetum purpureum* (Wan and Vasil 1996), *Poa pratensis* (Nielsen et al. 1993; Nielsen and Knudsen 1993) and *Zoysia japonica* (Inokuma et al. 1996; Table 5.1).

In bromegrass, *B. inermis*, only albino plantlets were regenerated from suspension cultures in an early study (Gamborg et al. 1970). The culture of corresponding cell suspension-derived protoplasts in media combining high sucrose and calcium led to sustained cell divisions, embryo formation and regeneration of green plantlets (Kao et al. 1973).

Embryogenic cell suspensions were initiated from calli induced on immature embryos and young inflorescences in *P. americanum* (Vasil and Vasil 1980), *P. maximum* (Lu et al. 1981) and *P. purpureum* (Vasil et al. 1983) for the isolation and culture of morphogenically competent protoplasts. The cultured protoplasts gave rise to embryogenic cell masses, embryoids and green plantlets. Plantlet formation was through somatic embryogenesis in all of these cases, and not after extensive callus growth preceding the organization of separate shoot and root meristems. However, it was not possible to establish soil-grown protoplast-derived plants in any of the cases (Vasil et al. 1983). Somatic embryos formed in callus cultures were easier to grow into plants than those developed from supension cultures or protoplasts (Vasil et al. 1983) despite suspension cultures of *P. maximum* and *P. purpureum* being comprised predominantly of normal diploid cells (Karlsson and Vasil 1986).

Following the procedures described earlier for *P. maximum* (Lu et al. 1981), protoplast cultures have been established in *P. miliaceum* (Heyser 1984). Although protoplast plating efficiencies of 0.4% were obtained, only callus and albino plantlets were recovered.

Suspension cultures of cocksfoot, *D. glomerata*, consisting of small embryogenic cell clusters have been developed from calli induced on basal sections of young leaves (Gray et al. 1984; Horn et al. 1988). The established cell suspensions contained numerous developing embryos and possessed few or no root primordia when casein hydrolysate was added to the culture medium. The somatic embryos germinated either in liquid or on solid medium lacking auxins, and the resulting plantlets were successfully transferred to soil (Gray et al. 1984). Cell suspension-derived protoplasts of *D. glomerata* cultured in an agarose-solidified medium showed plating efficiencies varying between 0.5 and 10% (Horn et al. 1988). The use of conditioned medium was found to increase the rate of cell colony formation from cultured cocksfoot protoplasts. The protoplast-derived colonies were embryogenic and formed green plantlets which were established in soil and grown until maturity (Horn et al. 1988).

Genotype screening using mature seeds has been performed for the initiation of embryogenic suspension cultures in redtop, *A. alba* (Asano and Sugiura 1990). Both green and albino plantlets were regenerated from the established cell suspensions. The average green:albino ratio was 19:1. These regenerable cell suspensions were used as a source of totipotent protoplasts. Protoplast yields of $1–2 \times 10^7$ protoplasts/g fresh weight suspension cells were obtained. These relatively high yields were achieved by subculturing donor cell suspensions with sucrose-free MS medium prior to protoplast isolation in order to reduce the large deposits of starch grains in the protoplasts, thus resulting in decreased protoplast lysis. Protoplast plating efficiencies varied between 0.5 and 0.6%. Green and albino plantlets developed from 50% of the protoplast-derived microcalli. Green:albino ratios of regenerants derived from protoplasts were similar to those obtained for corresponding suspension cultures (Asano and Sugiura 1990).

For creeping bentgrass, *A. palustris*, rapidly growing regenerable (64% regeneration frequency) cell suspensions have been established from seed-derived calli (Terakawa et al. 1992). The regeneration capacity of the suspension cells was maintained for more than 1 year. Protoplasts isolated from these cells and cultured in MS-based medium showed plating efficiencies of up to 0.36% when conditioned medium was used. Approximately 30% of the protocalli differentiated green plantlets, and the green:albino ratio was 27:1. Normal creeping bentgrass plants from protoplasts were successfully established in soil (Terakawa et al. 1992).

An improved response of colony formation has been observed for *Agrostis* protoplasts when cultured in higher agarose-concentration media (Asano et al. 1994). Protoplast plating efficiencies were greatly improved by increasing the agarose concentration. The use of 2.4% agarose increased the plating efficiency about 15- and 4-fold over the conventionally used agarose level (1.2%) in *A. alba* and *A. palustris*, respectively (Asano et al. 1994).

In apomictic dallisgrass, *P. dilatatum*, embryogenic cell suspensions have been successfully established from calli initiated from immature inflorescences

(Akashi and Adachi 1992) and mature seed-derived embryos (Legris et al., unpubl.). A 12-h pretreatment of suspension cells in liquid MS medium without sucrose and hormones significantly increased protoplast yield and colony formation. Plating efficiencies between 0.5 and 0.8% were obtained. Approximately 15% of the protoplast-derived calli formed plantlets on regeneration medium. Green plants only were regenerated (Akashi and Adachi 1992). In apomictic Bahiagrass, *P. notatum,* embryogenic cell suspensions have been established from seed-derived callus. Plant regeneration frequency of 28.6% was achieved when cell clusters derived from suspension cultures were transferred to regeneration medium (Akashi et al. 1993).

Plant regeneration from embryogenic suspension cultures has been reported for the minor millet, *P. scrobiculatum* (Nayak and Sen 1989). Over 60% of the plated suspension cells proliferated on solidified medium and produced embryogenic calli. Up to 90% of the embryogenic calli were able to regenerate plantlets. Regenerated plants were grown under field conditions and appeared morphologically normal (Nayak and Sen 1989).

In Kentucky bluegrass, *P. pratensis*, suspension cultures capable of regenerating green plants were established from embryogenic calli derived from either entire mature seeds or isolated embryos (Nielsen and Knudsen 1993). The regeneration frequencies from the established cell suspensions varied considerably: from 5 to >1000 plants/g fresh weight suspension cells. Cell suspension derived plants were transferred to soil and established under field conditions where they set seeds (Nielsen and Knudsen 1993). Plant regeneration from Kentucky bluegrass protoplasts isolated from one suspension culture was achieved and retained 10–16 months after establishment of the respective cell suspension (Nielsen et al. 1993). Protoplasts cultured in agarose beads in medium devoid of nurse or feeder cells achieved plating efficiencies between 0.004 and 1.5%. Well-formed somatic embryos developed from 0.4–2.7% of the protoplast-derived microcolonies. These somatic embryos were germinated and gave rise to green plants only (Nielsen et al. 1993). In an earlier report, albino plantlets of Kentucky bluegrass were exclusively regenerated from both suspension cells and corresponding protoplasts (van der Valk et al. 1988).

5.5 Summary and Conclusions

Recent achievements demonstrate that relatively large numbers of mature green plants that survived transfer to soil and were grown under greenhouse and field conditions could be reproducibly regenerated from embryogenic cell suspensions and corresponding protoplasts in different fescues and ryegrasses, as well as other grass species. This point has previously been the main constraint on the application of suspension- or protoplast-dependent genetic manipulation techniques for pasture and turf grass improvement.

A clear genotype dependence for obtaining callus cultures suitable for establishing embryogenic cell suspensions has been observed for the *Festuca* and *Lolium* species considered. Cultivars of these outcrossing pasture and turf grasses represent populations of highly heterozygous genotypes. On this basis, the selection of genotypes capable of producing embryogenic calli suitable for the initiation of corresponding single genotype-derived cell suspensions is a key prerequisite to obtaining a convenient source of regenerable target cells (suspension cells and protoplasts) for genetic manipulation. It is apparent that limited progress towards establishing reproducible systems to regenerate soil-grown plants from suspension cells and protoplasts in fescues and ryegrasses has been obtained from studies evaluating a wide range of media constituents. In contrast, the use of esentially similar, simple culture media and selected single genotype-derived embryogenic suspension cultures has allowed the reproducible regeneration of plants from suspension cells and protoplasts for a wide range of cultivars and genotypes in fescues and ryegrasses. This suggests that the morphogenetic competence of the cells, rather than the medium composition, holds the key to the successful culture of these graminaceous protoplasts.

The phenomenon of albinism from somatic tissue cultures is well known in different grass species, particularly with increasing age of cultures. However, green mature plants can be regenerated from these graminaceous protoplasts if they are isolated from young embryogenic cell suspensions before the ability of these to regenerate green plants has ceased. In addition, the use of cryopreservation for storage of young embryogenic suspension cultures in liquid nitrogen can further extend the useful life of established cell suspensions in fescues and ryegrasses, and enable protoplast culture experiments with specific genotypes to be performed repeatedly over a long period of time.

In summary, reproducible and efficient protoplast-to-plant regeneration systems in different *Festuca* and *Lolium* species have been established. These systems are based on:

1. Genotype screening of approximately 200–2000 seeds or embryos per cultivar for the identification of genotypes producing friable and highly embryogenic calli of single-seed origin.
2. Establishment of corresponding single genotype-derived highly embryogenic cell suspensions.
3. Cryopreservation of established embryogenic suspension cultures for their long-term availability.
4. Isolation of morphogenic protoplasts from young highly embryogenic cell suspensions.
5. Culture of protoplasts in a bead-type system including non-morphogenic nurse cells.

These plant regeneration systems represent a solid experimental basis for performing genetic manipulations at the cellular level. They will facilitate manipulations such as protoplast fusion for the production of somatic hybrids and cybrids (Chap. 6), direct gene transfer to protoplasts (Chap. 7) and gene transfer via biolistics or whisker-mediated transformation with embryogenic suspension cells (Chap. 8) for the generation of transgenic plants.

References

Akashi R, Adachi T (1992) Plant regeneration from suspension culture-derived protoplasts of apomictic dallisgrass (*Paspalum dilatatum* Poir.). Plant Sci 82: 219–225

Akashi R, Hashimoto A, Adachi T (1993) Plant regeneration from seed-derived embryogenic callus and cell suspension cultures of bahiagrass (*Paspalum notatum*). Plant Sci 90: 73–80

Altpeter F, Posselt UK (1994) Regeneration of fertile plants from cell-suspension-derived small-cell colonies of *Lolium perenne* (L.). Plant Breed 113: 327–330

Asano Y, Sugiura K (1990) Plant regeneration from suspension culture-derived protoplasts of *Agrostis alba* L. (Redtop). Plant Sci 72: 267–273

Asano Y, Ito Y, Sugiura K, Fujie A (1994) Improved protoplast culture of bentgrass (*Agrostis* L.) using a medium with increased agarose concentration. J Plant Physiol 143: 122–124

Creemers-Molenaar J, Van der Valk P, Loeffen JPM, Zaal MACM (1989) Plant regeneration from suspension cultures and protoplasts of *Lolium perenne* L. Plant Sci 63: 167–176

Creemers-Molenaar J, Loeffen JPM, Van Rossum M, Colijn-Hooymans CM (1992a) The effect of genotype, cold storage and ploidy level on the morphogenic response of perennial ryegrass (*Lolium perenne* L.) suspension cultures. Plant Sci 83: 87–94

Creemers-Molenaar J, Van Eeuwijk FA, Krens FA (1992b) Culture optimization for perennial ryegrass protoplasts. J Plant Physiol 139: 303–308

Dalton SJ (1988a) Plant regeneration from cell suspension protoplasts of *Festuca arundinacea* Schreb. (tall fescue) and *Lolium perenne* L. (perennial ryegrass). J Plant Physiol 132: 170–175

Dalton SJ (1988b) Plant regeneration from cell suspension protoplasts of *Festuca arundinacea* Schreb., *Lolium perenne* L. and *L. multiflorum* Lam. Plant Cell Tissue Organ Culture 12: 137–140

Dalton SJ, Thomas ID (1992) A statistical comparison of various factors on embryogenic proliferation, morphogenesis and regeneration in *Lolium temulentum* cell suspension colonies. Plant Cell Tissue Organ Culture 30: 15–29

Folling M, Madsen S, Olesen A (1995) Effect of nurse culture and conditioned medium on colony formation and plant regeneration from *Lolium perenne* protoplasts. Plant Sci 108: 229–239

Fournier D, Ghesquiere M, Poisson C (1996) Plant regeneration from cell suspension cultures of tetraploid tall fescue. Plant Cell Tissue Organ Culture 46: 165–168

Gamborg OL, Constabel F, Miller RA (1970) Embryogenesis and production of albino plants from cell cultures of *Bromus inermis*. Planta 95: 355–358

Gray DJ, Conger BV, Hanning GE (1984) Somatic embryogenesis in suspension and suspension-derived callus cultures of *Dactylis glomerata*. Protoplasma 122: 196–202

Heyser JW (1984) Callus and shoot regeneration from protoplasts of proso millet (*Panicum miliaceum* L.). Z Pflanzenphysiol 113: 293–299

Horn ME, Conger BV, Harms CT (1988) Plant regeneration from protoplasts of embryogenic suspension cultures of orchardgrass (*Dactylis glomerata* L.). Plant Cell Rep 7: 371–374

Inokuma C, Sugiura K, Cho C, Okawara R, Kaneko S (1996) Plant regeneration from protoplasts of Japanese lawngrass. Plant Cell Rep 15: 737–741

Jones MGK, Dale PJ (1982) Reproducible regeneration of callus from suspension culture protoplasts of the grass *Lolium multiflorum*. Z Pflanzenphysiol 105: 267–274

Kao KN, Gamborg OL, Michayluk MR, Keller WA, Miller RA (1973) The effects of sugars and inorganic salts on cell regeneration and sustained division in plant protoplasts. Colloques internationaux CNRS no 212, protoplastes et fusion de cellules somatiques végétales: 207–213

Kartha KK (1987) Advances in the cryopreservation technology of plant cells and organs. In: Green CE, Somers DA, Hackett WP, Biersboer DD (eds) Plant tissue and cell culture, Liss, New York, pp 447–458

Karlsson SB, Vasil IK (1986) Growth, cytology and flow cytometry of embryogenic cell suspension cultures of *Panicum maximum* Jacq. and *Pennisetum purpureum* Schum. J Plant Physiol 123: 211–227

Lu CY, Vasil V, Vasil IK (1981) Isolation and culture of protoplasts of *Panicum maximum* Jacq. (Guinea grass): somatic embryogenesis and plantlet formation. Z Pflanzenphysiol 104: 311–318

Morrish F, Vasil V, Vasil IK (1987) Developmental morphogenesis and genetic manipulation in tissue and cell cultures of the Gramineae. Adv Genet 24: 431–499

Müller AJ, Grafe R (1978) Isolation and characterization of cell lines of *Nicotiana tabacum* lacking nitrate reductase. Mol Gen Genet 161: 67–76

Murashige T, Skoog F (1962) A revised medium for rapid growth and bioassays with tobacco tissue culture. Physiol Plant 15: 473–497

Nayak P, Sen SK (1989) Plant regeneration through somatic embryogenesis from suspension cultures of a minor millet, *Paspalum scrobiculatum*. Plant Cell Rep 8: 296–299

Nielsen KA, Knudsen E (1993) Regeneration of green plants from embryogenic suspension cultures of Kentucky bluegrass (*Poa pratensis* L.). J Plant Physiol 141: 589–595

Nielsen KA, Larsen E, Knudsen E (1993) Regeneration of protoplast-derived green plants of Kentucky bluegrass (*Poa pratensis* L.). Plant Cell Rep 12: 537–540

Olesen A, Storgaard M, Madsen S, Andersen SB (1995) Somatic in vitro culture response of *Lolium perenne* L.: genetic effects and correlations with anther culture. Euphytica 86: 199–209

Potrykus I (1990) Gene transfer to cereals: an assessment. Bio/Technology 8: 535–542

Potrykus I, Shillito RD (1986) Protoplasts: isolation, culture, plant regeneration. Methods Enzymol 118: 549–578

Rajoelina SR, Alibert G, Planchon C (1990) Continuous plant regeneration from established embryogenic cell suspension cultures of Italian ryegrass and tall fescue. Plant Breed 104: 265–271

Roest S, Gilissen LJW (1989) Plant regeneration from protoplasts: a literature review. Acta Bot Neerl 38: 1–23

Spangenberg G (1995) Embryogenic cultures: a source of target cells for gene transfer. In: Potrykus I, Spangenberg G (eds) Gene transfer to plants. Springer, Berlin Heidelberg New York, pp 293–294

Spangenberg G, Wang ZY, Nagel J, Potrykus I (1994) Protoplast culture and generation of transgenic plants in red fescue (*Festuca rubra* L.). Plant Sci 97: 83–94

Stadelmann FJ, Boller B, Spangenberg G, Kölliker R, Messerli M, Wang ZY, Nösberger J (1997) Fertility and growth in the field of *Lolium perenne* and *Festuca rubra* plants regenerated from suspension cultured cells and protoplasts. Plant Breed 11 (in press)

Stadelmann FJ, Boller B, Spangenberg G, Kölliker R, Wang ZY, Potrykus I, Nösberger J (1998) Field performance of cell suspension-derived *Lolium perenne* L. regenerants and their progenies. Theor Appl Genet (in press)

Takamizo T, Suginobu KI, Ohsugi R (1990) Plant regeneration from suspension culture derived protoplasts of tall fescue (*Festuca arundinacea* Schreb.) of a single genotype. Plant Sci 72: 125–131

Terakawa T, Sato T, Koike M (1992) Plant regeneration from protoplasts isolated from embryogenic suspension cultures of creeping bentgrass (*Agrostis palustris* Huds.). Plant Cell Rep 11: 457–461

Vallés MP, Wang ZY, Montavon P, Potrykus I, Spangenberg G (1993) Analysis of genetic stability of plants regenerated from suspension cultures and protoplasts of meadow fescue (*Festuca pratensis* Huds.). Plant Cell Rep 12: 101–106

Van der Valk P, Zaal MACM, Creemers-Molenaar J (1988) Regeneration of albino plantlets from suspension culture derived protoplasts of *Poa pratensis* L. (Kentucky bluegrass). Euphytica S: 169–176

Vasil IK (1987) Developing cell and tissue culture systems for the improvement of cereal and grass crops. J Plant Physiol 128: 193–218

Vasil IK (1988) Progress in the regeneration and genetic manipulation of cereal crops. Bio/Technology 6: 397–402

Vasil V, Vasil IK (1980) Isolation and culture of cereal protoplasts, part 2: embryogenesis and plantlet formation from protoplasts of *Pennisetum americanum*. Theor Appl Genet 56: 97–99

Vasil V, Wang DY, Vasil IK (1983) Plant regeneration from protoplasts of napier grass (*Pennisetum purpureum* Schum.). Z Pflanzenphysiol 111: 233–239

Wan CH, Vasil IK (1996) Regeneration of plants from embryogenic callus, cell suspensions, protoplasts and cryopreserved cell suspension cultures of napiergrass (*Pennisetum purpureum* Schum.). J Plant Physiol 148: 718–726

Wang L, Wang X, Huang B (1996) Direct somatic embryogenesis and plant regeneration from single cell suspension cultures of *Elymus giganteus* Vahl. Plant Cell Rep 15: 865–868

Wang ZY, Vallés MP, Montavon P, Potrykus I, Spangenberg G (1993a) Fertile plant regeneration from protoplasts of meadow fescue (*Festuca pratensis* Huds.). Plant Cell Rep 12: 95–100

Wang ZY, Nagel J, Potrykus I, Spangenberg G (1993b) Plants from cell suspension-derived protoplasts in *Lolium* species. Plant Sci 94: 179–193

Wang ZY, Legris G, Nagel J, Potrykus I, Spangenberg G (1994) Cryopreservation of embryogenic cell suspensions in *Festuca* and *Lolium* species. Plant Sci 103: 93–106

Wang ZY, Legris G, Valles MP, Potrykus I, Spangenberg G (1995) Plant regeneration from suspension and protoplast cultures in the temperate grasses *Festuca* and *Lolium*. In: Terzi M, Cella R, Falavigna A (eds) Current issues in plant molecular and cellular biology. Kluwer Academic, Dordrecht, pp 81–86

Zaghmout OMF, Torello WA (1989) Somatic embryogenesis and plant regeneration from suspension cultures of red fescue. Crop Sci 29: 815–817

Zaghmout OMF, Torello WA (1990a) Isolation and culture of protoplasts from embryogenic suspension cultures of red fescue (*Festuca rubra* L.). Plant Cell Rep 9: 340–343

Zaghmout OMF, Torello WA (1990b) Somatic embryogenesis and plant regeneration from embryogenic suspension cultures of perennial ryegrass. In Vitro Cell Dev Biol 26: 419–424

Zaghmout OMF, Torello WA (1992a) Restoration of regeneration potential of long-term cultures of red fescue (*Festuca rubra* L.) by elevated sucrose levels. Plant Cell Rep 11: 142–145

Zaghmout OMF, Torello WA (1992b) Plant regeneration from callus and protoplasts of perennial ryegrass (*Lolium perenne* L.). J Plant Physiol 140: 101–105

Chapter 6

Somatic Hybridization

6.1 Introduction

Plant protoplasts may fuse when subjected immediately after isolation to specific experimental treatments such as polyethylene glycol (PEG) in the presence of high concentrations of calcium cations or electrical pulses after dielectrophoresis. The hybridization of different plant somatic cells after the fusogenic treatment of two distinct mixed populations of protoplasts is thus possible.

Fusion of plant protoplasts is non-specific and can therefore be carried out between any two species, irrespective of taxonomic relationships. In some cases, hybrid cells produced in this manner may subsequently be regenerated into entire hybrid plants. This process of non-sexual hybridization by protoplast fusion has been termed "parasexual hybridization" or "somatic hybridization", and the corresponding hybrid cells and hybrid plants obtained are referred to as "parasexual hybrids" or "somatic hybrids".

In "symmetric" protoplast fusion, the nuclear and cytoplasmic genophores of both parental protoplasts are combined in the fusion product. However, after fusion of protoplasts the segregation or complete loss of one of the parental cell nuclei may occur. In this case the hybrids obtained carrying the nuclear genes of one parent and the cytoplasmic genes of both parents or of the second parent only are called "cytoplasmic hybrids" or "cybrids". This term is also often used to describe "cytoplasmic heterozygotes", i.e. hybrids possessing alternative cytoplasmic genes from both parents.

A more or less extensive elimination of one of the parental nuclei can also be experimentally favoured in "asymmetric" protoplast fusions, e.g. by irradiation (with γ- or X-rays) of donor protoplasts prior to fusogenic treatment. Thus, depending on the genomic makeup of the recovered plants, symmetric and asymmetric somatic hybrids as well as cybrids can be generated (Schieder and Vasil 1980; Gleba and Sytnik 1984; Kumar and Cocking 1987). Methods for the selection and analysis of somatic hybrids and cybrids have been reviewed (Gleba and Shlumukov 1990; Medgyesy 1990). Somatic hybrid and cybrid plants have been regenerated from protoplast fusions in different plant families between different species, genera and tribes (Glimelius et al. 1991).

Somatic hybridization by protoplast fusion thus allows the production of novel combinations of parental genomes in higher plants. Plants having a nucleus originating from one parent and cytoplasm from the other, plants which are heterozygous for extranuclear genes, and plants bearing some extranuclear genes from one parent and the remaining extranuclear genes from the other serve as examples in this context. These gene combinations are either very difficult or impossible to generate by conventional sexual crossing (Gleba and Sytnik 1984).

Therefore, somatic hybridization by protoplast fusion shows promise in complementing conventional breeding programs of forage and turf grasses where: (1) the transfer of genes from wild relatives to cultivated plants is prevented by existing sexual crossing barriers; (2) the rapid combination of partial genomes from different sexually (in)compatible genetic origins in a predictable direction is needed to generate materials that could serve as bridges for the transfer of specific traits; and (3) the generation of cytoplasmic hybrids and/or the transfer of cytoplasmically encoded characters without numerous backcrosses is desired.

6.2 Somatic Hybridization in *Festuca* and *Lolium*

Due mainly to the difficulties faced over many years in the culture and regeneration of graminaceous monocot protoplasts (Vasil 1987), very few attempts have been made in the past to obtain somatic hybrid/cybrid plants in forage and turf grasses (Table 6.1). Only with the development of efficient protoplast-to-plant regeneration systems (Sect. 5.4) were opportunities to study somatic

Table 6.1. Somatic hybridization in forage and turf grasses

Plant species combination		Fusion product	References
Festuca arundinacea (tall fescue)	*Lolium multiflorum* (Italian ryegrass)	Symmetric somatic hybrid plants	Takamizo et al. (1991)
Festuca arundinacea (tall fescue)	*Lolium multiflorum* (Italian ryegrass)	Asymmetric somatic hybrid plants	Spangenberg et al. (1994)
Lolium perenne (perennial ryegrass)	*Lolium perenne* (perennial ryegrass)	Cybrid calli	Creemers-Molenaar et al. (1992)
Pennisetum americanum (pearl millet)	*Panicum maximum* (guinea grass)	Somatic hybrid calli	Ozias-Atkins et al. (1986)
Pennisetum americanum (pearl millet)	*Saccharum officinarum* (sugarcane)	Somatic hybrid calli	Tabaeizadeh et al. (1986)
Pennisetum americanum (pearl millet)	*Triticum monococcum* (einkorn)	Somatic hybrid calli	Vasil et al. (1988)

hybridization and cybridization opened in fescues and ryegrasses (Takamizo and Spangenberg 1994; Spangenberg et al. 1995).

For decades, *Festuca-Lolium* hybrids have been produced by conventional crossing in order to combine the agronomically desirable traits from species belonging to these genera (Crowder 1953). However, sexual hybrids between *L. multiflorum* and *F. arundinacea* have been readily obtained only unidirectionally (Italian ryegrass x tall fescue), while the reciprocal cross has been shown to be extremely difficult (Eizenga and Buckner 1986). In addition, chromosomal instabilities and the poor female fertility observed in the early amphiploid *Festulolium* plants derived from these octoploid sexual hybrids have so far severely limited the release of valuable materials from these crossings.

The recovery of flowering somatic hybrid plants derived from graminaceous monocot species belonging to different genera was reported for the first time in the combination *F. arundinacea* (+) *L. multiflorum* (Takamizo et al. 1991). Asymmetric somatic hybrid plants have also been obtained for the same intergeneric protoplast fusion combination (Spangenberg et al. 1994). Protoplasts isolated from morphogenic cell suspensions of *F. arundinacea* (Fig. 6.1A) and non-morphogenic suspension cultures of *L. multiflorum* (Fig. 6.1B) were used in these studies (Takamizo and Spangenberg 1994). A selection scheme based on metabolically inactivated totipotent protoplasts of tall fescue and unirradiated or X-ray irradiated non-morphogenic protoplasts of Italian ryegrass proved suitable for the enrichment of protoplast fusion products for the generation of symmetric and asymmetric somatic hybrids, respectively (Spangenberg et al. 1995). Treatment of tall fescue protoplasts with iodoacetamide (10 mM for 15 min) was reported to completely inhibit colony formation (Takamizo et al. 1991; Spangenberg et al. 1994). X-ray irradiation of Italian ryegrass protoplasts in the range of 50–500 Gy revealed that doses of 100 Gy and more fully abolished colony formation (Spangenberg et al. 1994).

Electrofusion experiments performed using unirradiated (for symmetric fusions) or 10–500 Gy X-ray irradiated (for asymmetric fusions) Italian ryegrass protoplasts (Fig. 6.1C) led to colony formation within 3–4 weeks in bead-type culture (Fig. 6.1D, E) and more than 60 calli that differentiated green shoots (Fig. 6.1F; Spangenberg et al. 1994, 1995). Based on the tight selection scheme used for enrichment of fusion products, these regenerants were preliminarily identified as presumptive symmetric and asymmetric somatic hybrid clones (Spangenberg et al. 1995). Plant habitus and leaf morphology were found to vary among these independent symmetric and asymmetric tall fescue (+) Italian ryegrass somatic hybrid plants grown until maturity under greenhouse conditions (Spangenberg et al. 1994; Takamizo and Spangenberg 1994). Some plants showed an intermediate character when compared to parental plants, while others more closely resembled tall fescue (Fig. 6.1G). In other cases, chimeric clones revealing plant sectors with leaf morphology and plant habitus similar to Italian ryegrass (Fig. 6.1H), indicative of the occurrence of mitotic instabilities, were described (Takamizo and

A
B
C
D
E
F
G
H
I
J
K

Spangenberg 1994; Spangenberg et al. 1995). Some of these plants were brought to flower and produced inflorescences with protruding anthers which contained viable pollen (Fig. 6.1I). Crosses of unstable clonal plants with Italian ryegrass pollen led to seed formation (Fig. 6.1J; Takamizo and Spangenberg 1994). Agronomical evaluation under field conditions of representative primary intergeneric somatic hybrid *Festulolium* plants obtained from symmetric and asymmetric fusion combinations (Fig. 6.1K; Spangenberg et al. 1995) and progenies recovered from crosses of unstable somatic *Festulolium* hybrids (T. Takamizo, pers. comm.) are in progress.

The analysis of the nuclear and organellar composition of these symmetric and asymmetric somatic *Festulolium* plants has been reported (Takamizo et al. 1991; Spangenberg et al. 1994, 1995). Quantitative dot blot and Southern hybridizations using interdispersed repetitive DNA probes specific for Italian ryegrass and tall fescue (Sect. 9.3) proved useful for studying the genomic makeup of both sexual and somatic *Festulolium* plants (Takamizo et al. 1991; Pérez-Vicente et al. 1992; Spangenberg et al. 1994). Using these techniques, the presence of Italian ryegrass DNA in the nuclear genome of all primary regenerants from symmetric fusions could be demonstrated, and hybridization signals estimated to be approximately one genome equivalent of *L. multiflorum* per haploid genome of *F. arundinacea* were found (Takamizo et al. 1991; Spangenberg et al. 1995). These results confirmed the true nuclear hybrid nature of all green plants regenerated from symmetric fusions, and the tightness of the iodoacetamide treatment used for the metabolic inactivation of morphogenic protoplasts of *F. arundinacea*.

A similar analysis was based on quantitative dot blot hybridizations and determined the nuclear composition of asymmetric somatic *Festulolium* hybrids derived from fusion experiments using Italian ryegrass donor protoplasts irradiated with X-rays of 10, 25, 50, 100 and 250 Gy (Fig. 6.2A, B) as well as 500 Gy (Fig. 6.2C–E; Spangenberg et al. 1994, 1995). In these reports, a

Fig. 6.1A–K. Recovery of symmetric and asymmetric somatic hybrid plants between *Festuca arundinacea* Schreb. and *Lolium multiflorum* Lam. **A** Embryogenic cell suspension of *F. arundinacea* used for isolation of totipotent protoplasts. **B** Non-morphogenic cell suspension of *L. multiflorum* used for isolation of protoplasts. **C** Protoplast mix after fusion of iodoacetamide (IOA) inactivated protoplasts of *F. arundinacea* and non-morphogenic protoplasts of *L. multiflorum* isolated from **A** and **B**, respectively. **D** Multiple divisions of putative fusant from asymmetric fusion between IOA-inactivated tall fescue protoplasts and 250-Gy X-ray irradiated Italian ryegrass protoplasts 2 weeks after fusion. **E** Putative somatic hybrid colonies obtained from fusion experiments between tall fescue protoplasts and 500-Gy irradiated Italian ryegrass protoplasts after 4 weeks in bead-type culture. **F** Regenerating clone from **E** 2 months after protoplast fusion. **G** Greenhouse-growing mature somatic hybrid *Festulolium* plants obtained from symmetric and asymmetric protoplast fusions. **H** Morphology of clonally unstable somatic hybrid plant (*right side* of the plant is morphologically similar to Italian ryegrass). **I** Acetocarmine-stained pollen of somatic hybrid *Festulolium* plant shown in **H**. **J** Inflorescence with protruding anthers of the plant shown in **H**. **K** Seeds developing on symmetric somatic hybrid *Festulolium* plant after pollination with Italian ryegrass pollen. (Spangenberg et al. 1995)

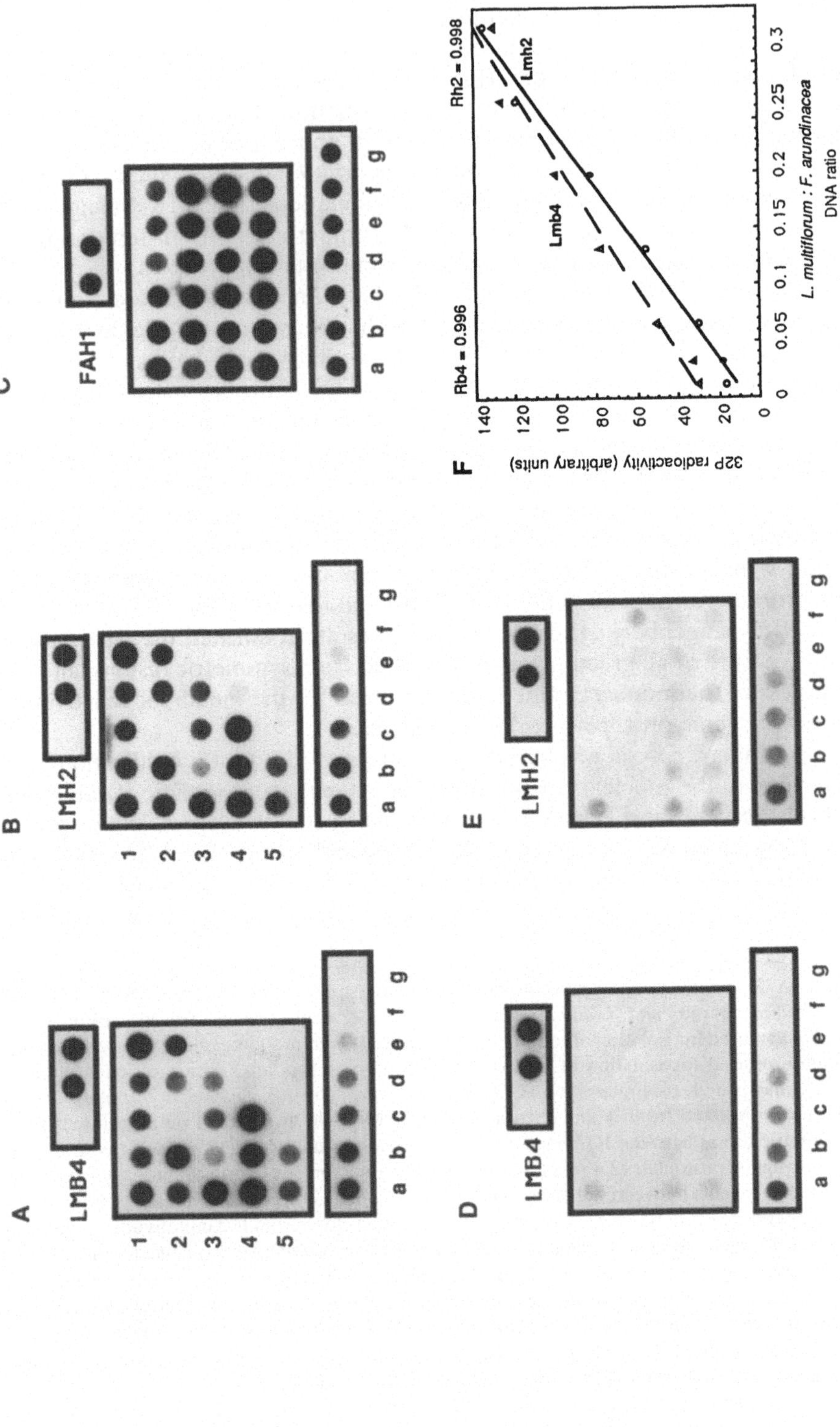
A
LMB4
B
LMH2
C
FAH1
D
LMB4
E
LMH2
F
Rb4 = 0.996
Rh2 = 0.998
Lmb4
Lmh2
32P radioactivity (arbitrary units)
L. multiflorum : F. arundinacea
DNA ratio

concentration series of parental DNAs was used to make a calibration plot of the radioactivity per dot in relation to the amount of DNA from one species and for mixes representing different ratios of genome equivalents (Fig. 6.2F; Pérez-Vicente et al. 1992; Spangenberg et al. 1994, 1995). With these calibration plots for two Italian ryegrass probes and one tall fescue species-specific probe (Pérez-Vicente et al. 1992), the amount of Italian ryegrass and tall fescue DNA per dot could be estimated and the fraction of nuclear DNA of the asymmetric somatic hybrids derived from Italian ryegrass determined (Spangenberg et al. 1994, 1995). All analyzed regenerating clones derived from 10- to 250-Gy fusion products (Fig. 6.2A, B) contained Italian ryegrass nuclear DNA but showed a large variation in the contribution of *L. multiflorum* DNA to their genomes (Fig. 6.3). These asymmetric somatic hybrid clones showed either no or limited donor genome elimination, thus being almost symmetric, or retained even less than 5% of the *L. multiflorum* genome, thus being highly asymmetric (Spangenberg et al. 1994, 1995). Some asymmetric somatic *Festulolium* hybrids that were recovered after irradiation with doses as different as 25 and 250 Gy retained comparable amounts of donor nuclear DNA, whereas the degree of asymmetry found in independent clones within each dose varied to a greater extent (Figs. 6.2A, B and 6.3) (Spangenberg et al. 1995). Analysis of the asymmetric somatic hybrid clones obtained from donor protoplasts irradiated with 500 Gy revealed extensive (more than 85%) and similar elimination of the Italian ryegrass nuclear genome for both Italian ryegrass-specific probes tested (Fig. 6.2D, E). For some of the 500-Gy asymmetric somatic hybrid clones analyzed, no *L. multiflorum* DNA above background was detectable (Fig. 6.3). Estimates from dot blots hybridized with a tall fescue-specific repetitive sequence as probe (Fig. 6.2C) indicated the presence of a complete or near complete chromosome set of the recipient *F. arundinacea* in most asymmetric somatic hybrid clones analyzed (Spangenberg et al. 1995). These results suggested that irradiation of donor cells prior to fusion led to unidirectional species-specific

Fig. 6.2A–F. Dot blot analysis of *F. arundinacea* (+) *L. multiflorum* asymmetric somatic hybrids. Quantitative dot blot analysis using *L. multiflorum*- (*LMH2*, *LMB4*) and *F. arundinacea*- (*FAH1*) specific sequences of asymmetric somatic hybrid clones (independent regenerating calli and plants) obtained from independent fusion products between recipient protoplasts of *F. arundinacea* and donor protoplasts of *L. multiflorum*. **A,B** Irradiation at *1* 10 Gy; *2* 25 Gy; *3* 50 Gy; *4* 100 Gy; and *5* 250 Gy. *a–g* Decreasing amounts of *L. multiflorum* DNA, as described in **F**. **C–E** Irradiation at 500 Gy. *Uppermost row* of dots represents (from *left* to *right*): 7.5 μg DNA of *F. arundinacea*, mix of 2.5 μg DNA of *L. multiflorum* and 7.5 μg DNA of *F. arundinacea*, and 2.5 μg DNA of *L. multiflorum*. **F** Calibration plots of the radioactivity per dot for decreasing amounts of *L. multiflorum* DNA: *a* 2.5 μg; *b* 2.0 μg; *c* 1.5 μg; *d* 1.0 μg; *e* 0.5 μg; *f* 0.25 μg; and *g* 0.1 μg in a constant (7.5 μg) background of *F. arundinacea* DNA (*bottom row* of **A–E**). Correlation coefficients (*R*) are given for data from hybridizations with *L. multiflorum*-specific sequences LMB4 (*Rb4*, plot *Lmb4*) and LMH2 (*Rh2*, plot *Lmh2*). *Open circles* refer to data obtained for hybridizations with the *L. multiflorum*-specific sequence LMB4; *triangles* for hybridizations with the *L. multiflorum*-specific sequence LMH2. (Spangenberg et al. 1994)

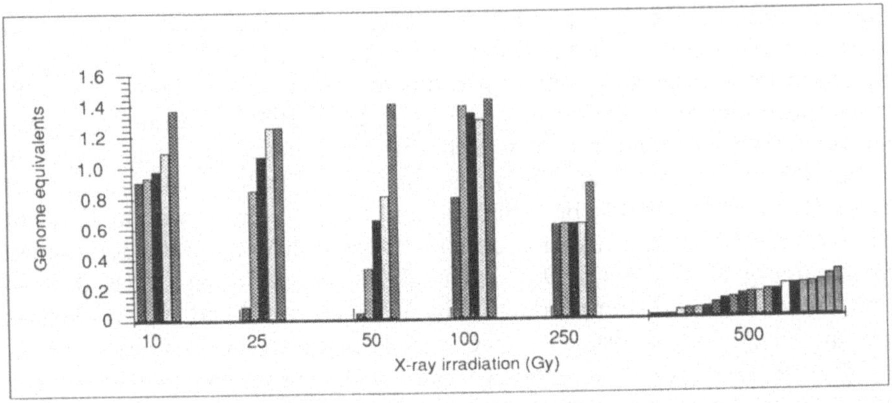

Fig. 6.3. Genome composition of asymmetric *F. arundinacea* (+) *L. multiflorum* somatic hybrids. Quantitative dot blots of asymmetric somatic *Festulolium* clones (*columns* correspond to independent regenerating calli and plants) obtained from different doses (10–500 Gy) of X-ray irradiation of donor protoplasts were hybridized with the *L. multiflorum* specific repetitive sequences LMH2 and LMB4. Genomic composition of independent clones based on mean values of data obtained from hybridization with both probes is shown as genome equivalents of *L. multiflorum* per genome equivalent of *F. arundinacea.* (Spangenberg et al. 1994)

elimination of *L. multiflorum* chromosomes, but no strict correlation between the level of species-specific *Lolium* genome elimination and the radiation dose used was apparent.

Information confirming the chromosomal composition and relative genomic contribution of both parents to the primary somatic *Festulolium* hybrid plants recovered was obtained from chromosome counts and a preliminary analysis by in situ hybridization to metaphase chromosomes using two *L. multiflorum*-specific repetitive DNA sequences (Spangenberg et al. 1994, 1995). However, the meiotic behaviour of these symmetric and asymmetric intergeneric *F. arundinacea* (+) *L. multiflorum* somatic hybrid plants remains to be studied.

The organellar composition of the intergeneric *F. arundinacea* (+) *L. multiflorum* symmetric and asymmetric somatic hybrids obtained was inferred from restriction fragment length polymorphism (RFLP) patterns obtained with different mitochondrial (mt) and chloroplast (cp) DNA specific heterologous gene probes (Takamizo et al. 1991; Spangenberg et al. 1994, 1995; Takamizo and Spangenberg 1994). Depending on the hybridization probe and the somatic *Festulolium* clone considered, the analysis of mitochondria type revealed parental-like, additive and novel (involving the absence of parental-like bands and/or the presence of non-parental bands) patterns. Representative results from this analysis with the maize *atpA* hybridization probe are shown for a whole set of *Festulolium* clones regenerated from symmetric (0 Gy) and asymmetric (10–500 Gy) protoplast fusions in Fig. 6.4 (Spangenberg et al. 1995). Additive patterns were preferentially found in the

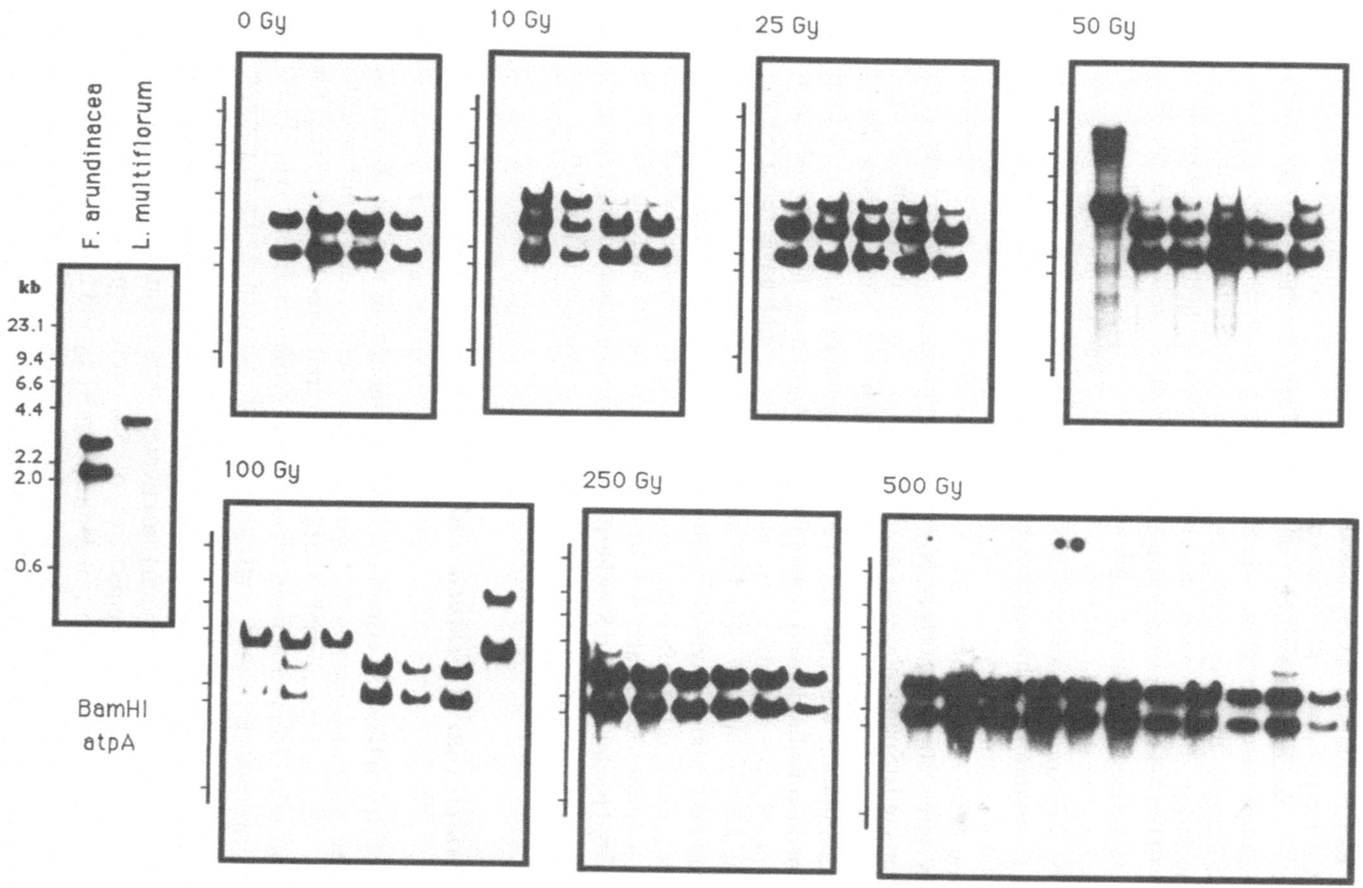

Fig. 6.4. RFLP analysis of organellar composition of symmetric and asymmetric *F. arundinacea* (+) *L. multiflorum* somatic hybrids. Analysis performed with mitochondrial gene specific probes *atpA* for symmetric (*0 Gy*) and asymmetric (*10–500 Gy*) somatic hybrid clones (independent regenerating calli and plants) obtained from fusions between iodoacetamide inactivated *F. arundinacea* protoplasts and unirradiated or X-ray irradiated *L. multiflorum* protoplasts. DNA samples were digested with *Bam*HI. The *left-most blot* shows corresponding hybridization patterns for the parental forms of *F. arundinacea* and *L. multiflorum*. (Spangenberg et al. 1995)

mtDNA RFLP analysis of symmetric and <50-Gy asymmetric clones. However, concomitantly with extensive nuclear genome elimination of the donor, *F. arundinacea* patterns became predominant for the >100-Gy asymmetric *Festulolium* clones (Fig. 6.4; Spangenberg et al. 1994, 1995). In an analogous manner, the chloroplast type of *F. arundinacea* (+) *L. multiflorum* symmetric and asymmetric somatic hybrids has been studied (Spangenberg et al. 1994, 1995; Takamizo and Spangenberg 1994). *F. arundinacea* cpDNA patterns have been reported in four independent symmetric somatic hybrid plants analyzed in detail (Takamizo et al. 1991; Takamizo and Spangenberg 1994). The corresponding analysis of asymmetric somatic *Festulolium* clones revealed *L. multiflorum* chloroplasts only in two cases, while in other cases mixes of both chloroplast types were still present. However, for the asymmetric somatic *Festulolium* plants derived from fusions using donor protoplasts irradiated with 250- and 500-Gy X-rays, *F. arundinacea* chloroplasts were mainly detected (Spangenberg et al. 1995). A preferential transmission towards recipient-type organelles occurring concomitantly with extensive donor genome elimination was thus apparent (Spangenberg et al. 1994). A final assessment of possibly biased organellar transmission in the *F. arundinacea* (+) *L. multiflorum* asymmetric somatic hybrids would have required the analysis of a larger number of independent regenerated plants.

Asymmetric somatic hybrid *Festulolium* plants have been obtained from protoplast fusions in the combination *F. rubra* (+) *L. perenne* (Legris 1996). The true asymmetric somatic hybrid nature of these regenerants has been demonstrated by dot blot and Southern hybridization analyses using repetitive DNA sequences specific for red fescue and Italian ryegrass (Chap. 10).

6.3 Cybridization in *Festuca* and *Lolium*

In addition to studies on symmetric and asymmetric protoplast fusions in intergeneric *Festuca-Lolium* combinations, experiments aimed at intraspecific transfer of the cytoplasmic male sterility (CMS) trait have been reported for *L. perenne* (Creemers-Molenaar et al. 1992). Donor-recipient type PEG-mediated protoplast fusions were carried out between 15-krad gamma-irradiated protoplasts from the cell line Lp9A (CMS *L. perenne* breeding line B200 as donor) and protoplasts from the cell line Lp25 (male fertile *L. perenne* breeding line Jon401 as recipient) inactivated with 1–7 m*M* iodoacetamide. Homo- and heterofusion frequencies of 5% were reported (Creemers-Molenaar et al. 1992). A sample of 25 presumptive cybrid calli was characterized with respect to mitochondria type and nuclear composition. The corresponding analysis of phosphoglucoisomerase isozyme patterns and ploidy level determination based on flow cytometry suggested the retention of nuclear DNA from the recipient in all cases. However, evidence indicative of the presence of some nuclear DNA from the donor was also obtained

(Creemers-Molenaar et al. 1992). Determination of the mitochondria type in the putative cybrid calli was based on RFLP patterns using two mitochondrial gene specific probes. These *atp6* and *cox1* probes had previously been shown to discriminate between male fertile and CMS perennial ryegrass mtDNA samples (Rouwendal et al. 1992). Parental-like, additive and novel patterns were found in this RFLP analysis. Furthermore, extensive qualitative and quantitative variation in mtDNA patterns from the cybrid calli analyzed was observed, which was indicative of fusion-induced inter- or intramolecular mitochondrial recombination (Creemers-Molenaar et al. 1992). However, in this work, non-morphogenic cell suspensions were used as a source of recipient protoplasts, thus a priori precluding opportunities for the recovery of proven cybrid plants and for a demonstration of intraspecific transfer of the CMS trait in perennial ryegrass.

Nevertheless, in recent years, the techniques required for the production of symmetric and asymmetric somatic hybrids as well as cybrids have been established for representative combinations in the *Festuca-Lolium* complex. In addition, progress has been made in setting up the methodical basis for defined cybridizations mediated by protoplast-cytoplast fusions, gametosomatic fusions and in vitro fertilization by fusion of isolated sperm and egg cells, particularly for perennial ryegrass. Along these lines, Van Ark et al. (1992) described a protocol for the efficient isolation of cytoplasts (enucleated cells) from protoplasts of *L. perenne* by centrifugation in discontinuous sucrose/mannitol gradients. Cytoplast yields of up to 45% of the perennial ryegrass protoplasts loaded onto the gradient were achieved, and the recovered cytoplast fractions contained only 2–8% contaminating protoplasts (Van Ark et al. 1992). This procedure should allow for cytoplast preparations of high purity to be obtained in sufficient quantities for mass fusion of protoplast-cytoplasts for directed cybridizations aimed, for example, at the intraspecific transfer of CMS in perennial ryegrass. These cytoplast-protoplast fusions could make it possible to avoid the retention of some donor nuclear DNA in the respective fusion products, a constraint observed in the work reported by Creemers-Molenaar et al. (1992), and could thus contribute to the generation of true cybrids after the exclusive transfer of the CMS-determining cytoplasm from the donor.

In addition to a protocol available for the large-scale preparation of pure cytoplasts in *L. perenne* (Van Ark et al. 1992), procedures for the isolation of viable sperm cells (Van der Maas et al. 1993a) and viable egg cells (Van der Maas et al. 1993b) have been reported. A method for the release of sperm cells from pollen grains of perennial ryegrass by osmotic shock has been partially optimized, and an isolation medium adjusted to pH 6.0 and 360 mOsmol/kg H_2O osmolarity was found to be best (Van der Maas et al. 1993a). The addition of 10 m*M* vitamin C to the isolation medium and 6.5% fetal calf serum to the storage medium allowed for an increase in sperm cell yields from 5 to 12% and an increase in storage time for retaining 50% of the sperm cell population viable from 10 to 60 h (Van der Maas et al. 1993a). Detailed cytological studies confirmed that these isolated populations of sperm cells from perennial ryegrass pollen grains had a high degree of purity, exhibited moderate cytological variability, and lacked cellulose and callose, thus representing true protoplasts (Van der Maas et al. 1993a, 1994).

By a combination of enzymatic maceration and micromanipulation, the isolation of viable egg cells from ovules of *L. perenne* has been possible (Van der Maas et al. 1993b). The yield of isolated egg cells was estimated at 10% of the treated ovules. Since the number of sperm cells (Van der Maas et al. 1993a) and cytoplasts (Van Ark et al. 1992) that can be isolated per experiment may be large, gametosomatic fusions [male gametes (+) somatic protoplast or cytoplast] are now conceivable based on mass fusion procedures in *L. perenne*. In contrast, in spite of the relatively high yields reported for the preparation of egg cells in perennial ryegrass (Van der Maas et al. 1993b), the corresponding overall number of egg cells obtained per experiment seemed too small to be useful in mass fusion experiments. With the development of the microfusion technique reviewed by Spangenberg (1994) and its proven utility for the in vitro fertilization in plants mediated by one-to-one electrofusion of egg cells and sperm cells, as convincingly demonstrated for maize (Kranz and Lörz 1993), analogous opportunities should become available for perennial ryegrass.

6.4 Somatic Hybridization in Other Grasses

In cereals and grasses, interspecific and intergeneric somatic hybridizations have been reported for different symmetric protoplast fusion combinations (Ozias-Akins et al. 1986; Tabaeizadeh et al. 1986; Terada et al. 1987; Hayashi et al. 1988; Vasil et al. 1988). In many instances, protoplasts from a dicot species have also been used as a fusion partner (Kao and Michayluk 1974; Kao et al. 1974; Dudits et al. 1976; Niizeki et al. 1985; Sala et al. 1985; Kisaka et al. 1994). In the early reports, evidence of heterokaryon formation was provided following the PEG-mediated fusion of mesophyll protoplasts of maize and barley with suspension cell-derived protoplasts of soybean (Kao and Michayluk 1974; Kao et al. 1974). Some of these putative fusion products, which were identified by the presence of chloroplasts and colorless cytoplasm, underwent few mitotic divisions. In analogous experiments, Dudits et al. (1976) identified heterokaryons from fusions of suspension cell-derived protoplasts of carrot with barley mesophyll protoplasts on the basis of differential nuclear staining. However, in all of these cases sustained cell divisions were obtained, and due to the lack of selection systems, the putative somatic hybrid cells were lost (Kao and Michayluk 1974; Kao et al. 1974; Dudits et al. 1976).

Somatic hybrid cell lines were obtained following PEG-mediated fusion of protoplasts from suspension cultured cells of rice and carrot (Sala et al. 1985) and after fusion of callus protoplasts from rice and soybean (Niizeki et al. 1985). The retention of a minor fraction of rice nuclear DNA and rapid elimination of rice chromosomes could be proven in these somatic hybrid cell lines, respectively. The recovery of highly asymmetric somatic hybrid plants between rice and carrot has also been reported (Kisaka et al. 1994). In this case, the morphology of the regenerated plants closely resembled that of the parental carrot plants, but the molecular characterization performed on the basis of peroxidase isozyme patterns and Southern hybridization analyses using nuclear, chloroplast and mitochondrial gene-specific probes confirmed the somatic hybrid nature of some of these regenerants.

Protoplast fusions between two different graminaceous monocots have also been reported (Ozias-Akins et al. 1986; Tabaeizadeh et al. 1986; Vasil et al. 1988). Metabolically inactivated protoplasts from a non-morphogenic S-2-aminoethyl-L-cysteine (AEC) cell line of pearl millet (*Pennisetum americanum*) were fused with protoplasts isolated from cell suspension cultures of sugarcane (*Saccharum officinarum*; Tabaeizadeh et al. 1986), guinea grass (*Panicum maximum*; Ozias-Akins et al. 1986) and einkorn (*Triticum monococcum*; Vasil et al. 1988). Selection with AEC allowed for the enrichment of presumptive somatic hybrid colonies. The true somatic hybrid nature of the calli obtained was demonstrated by isozyme patterns and Southern hybridization analysis using a maize ribosomal DNA probe. In spite of somatic embryos formed in hybrid calli from the pearl millet and guinea grass combination, no somatic hybrid plants were recovered in any of these combinations (Vasil 1987).

Somatic hybridization in another intergeneric graminaceous combination, namely rice (*O. sativa*) and barnyard grass (*Echinochloa oryzicola*), has been demonstrated (Terada et al. 1987). Selection of hybrids was based on the inability of barnyard grass protoplasts to divide and on the iodoacetamide inactivation of the rice protoplasts. A total of 166 hybrid calli were obtained from the electrofused protoplasts. Their somatic hybrid character was demonstrated by isozyme pattern analyses and chromosome counts. Out of these somatic hybrid calli, 44 formed shoots and 9 developed into in vitro-growing plantlets. The recovery of adult plants, however, failed (Terada et al. 1987). In contrast, viable somatic hybrid plants have been recovered for a few interspecific combinations in the genus *Oryza*, namely *O. sativa* (+) *O. eichingeri*, *O. sativa* (+) *O. officinalis*, *O. sativa* (+) *O. brachyantha*, and *O. sativa* (+) *O. perrieri* (Hayashi et al. 1988). Similarly, the generation of mature cybrid plants in the Poaceae seems restricted to rice so far and is mainly aimed at the intraspecific transfer of the CMS trait (Akagi et al. 1989; Kyozuka et al. 1989; Yang et al. 1989).

6.5 Summary and Conclusions

Genotypically and phenotypically different somatic *Festulolium* hybrid plants have been regenerated from symmetric and asymmetric protoplast fusions between *F. arundinacea* and *L. multiflorum*. Cloned interdispersed repetitive

DNA sequences specific to tall fescue and Italian ryegrass proved useful tools for analyzing the genomic composition of these somatic hybrids. X-ray irradiation of the donor protoplasts favoured unidirectional elimination of most or a few of the donor chromosomes in the somatic hybrid plants. These results demonstrate that for a model combination in the *Festuca-Lolium* complex, asymmetric protoplast fusion allows for directed one-step partial nuclear genome transfer in an intergeneric combination of grasses. These procedures show potential for facilitating limited alien gene transfer between sexually (in)compatible species to complement or enhance conventional wide hybridization programs in forage and turf grasses. However, the evaluation of fertility and meiotic stability of asymmetric somatic hybrids obtained in different close and remote combinations should allow this potential to be better assessed.

In addition, protocols for cybridization and isolation of sperm and egg cells have been partially optimized in *L. perenne*. Thus new opportunities for modifying organellar composition and generating novel nuclear-organellar interactions are now available for this species. It is expected that with some additional effort, similar protocols could be worked out for other forage and turf grasses. The recent advances in the regeneration of fertile plants from protoplasts in forage and turf grasses (Chap. 5), combined with the available methodical basis of protoplast fusion technology described here, stimulate hopes for future applications of these genetic manipulation techniques in corresponding breeding programs.

References

Akagi H, Sakamoto M, Negishi T, Fujimura T (1989) Construction of rice cybrid plants. Mol Gen Genet 215: 501–506

Creemers-Molenaar J, Hall RD, Krens FA (1992) Asymmetric protoplast fusion aimed at intraspecific transfer of cytoplasmic male sterility (CMS) in *Lolium perenne* L. Theor Appl Genet 84: 763–770

Crowder LV (1953) Interspecific and intergeneric hybrids of *Festuca* and *Lolium*. J Hered 44: 195–203

Dudits D, Kao KN, Constabel F, Gamborg OL (1976) Fusion of carrot and barley protoplasts and division of heterokaryocytes. Can J Genet Cytol 18: 263–269

Eizenga GC, Buckner RC (1986) Cytological and isozyme evaluation of tall fescue × Italian ryegrass hybrids. Plant Breed 97: 340–344

Gleba YY, Shlumukov LR (1990) Selection of somatic hybrids. In: Dix PJ (ed) Plant cell line selection. Procedures and applications. VCH, Weinheim, pp 257–286

Gleba YY, Sytnik KM (1984) Protoplast fusion. Genetic engineering in higher plants. In: Shoeman R (ed) Monographs on theoretical and applied genetics, vol 8. Springer, Berlin Heidelberg New York, 220 pp

Glimelius K, Fahlesson J, Landgren M, Sjodin C, Sundberg E (1991) Gene transfer via somatic hybridization in plants. TIBTECH 9: 24–30

Hayashi Y, Kyozuka J, Shimamoto K (1988) Hybrids of rice (*Oryza sativa* L.) and wild *Oryza* species obtained by cell fusion. Mol Gen Genet 214: 6–10

Kao KN, Michayluk MR (1974) A method for high-frequency intergeneric fusion of plant protoplasts. Planta 115: 355–367

Kao KN, Constabel F, Michayluk MR, Gamborg OL (1974) Plant protoplast fusion and growth of intergeneric hybrid cells. Planta 120: 215–227

Kisaka H, Lee H, Kisaka M, Kanno A, Kang K, Kameya T (1994) Production and analysis of asymmetric hybrid plants between monocotyledon (*Oryza sativa* L.) and dicotyledon (*Daucus carota* L.). Theor Appl Genet 89: 365–371

Kranz E, Lörz H (1993) In vitro fertilization with isolated, single gametes results in zygotic embryogenesis and fertile maize plants. Plant Cell 5: 739–746

Kumar A, Cocking EC (1987) Protoplast fusion: a novel approach to organelle genetics in higher plants. Am J Bot 74: 1289–1303

Kyozuka J, Kaneda T, Shimamoto K (1989) Production of cytoplasmic male sterile rice (*Oryza sativa* L.) by cell fusion. Bio/Technology 7: 1171–1174

Legris G (1996) Asymmetric somatic hybridization between red fescue (*Festuca rubra*) and perennial ryegrass (*Lolium perenne*). PhD Diss, Swiss Federal Institute of Technology, Zürich

Medgyesy P (1990) Selection and analysis of cytoplasmic hybrids. In: Dix PJ (ed) Plant cell line selection. Procedures and applications. VCH, Weinheim, pp 287–316

Niizeki M, Tanaka M, Akada S, Hirai A, Saito KI (1985) Callus formation of somatic hybrid of rice and soybean and characteristics of the hybrid callus. Jpn J Genet 60: 81–92

Ozias-Akins P, Ferl RJ, Vasil IK (1986) Somatic hybridization in the Gramineae: *Pennisetum americanum* (L.) K. Schum. (pearl millet) + *Panicum maximum* Jacq. (guinea grass). Mol Gen Genet 203: 365–370

Pérez-Vicente R, Petris L, Osusky M, Potrykus I, Spangenberg G (1992) Molecular and cytogenetic characterization of repetitive DNA sequences from *Lolium* and *Festuca*: applications in the analysis of *Festulolium* hybrids. Theor Appl Genet 84: 145–154

Rouwendal GJA, Creemers-Molenaar J, Krens FA (1992) Molecular aspects of cytoplasmic male sterility in perennial ryegrass (*Lolium perenne* L.) mtDNA and RNA differences between plants with male-sterile and fertile cytoplasm and restriction mapping of their *atp6* and *coxI* homologous regions. Theor Appl Genet 83: 330–336

Sala C, Biasini G, Morandi C, Nielsen E, Parisi B, Sala F (1985) Selection and nuclear DNA analysis of cell hybrids between *Daucus Carota* and *Oryza sativa*. J Plant Physiol 118: 409–419

Schieder O, Vasil IK (1980) Protoplast fusion and somatic hybridization. In: Vasil IK (ed) Perspectives in plant cell and tissue culture. Int Rev Cytol [Suppl] 11B: 21–46

Spangenberg G (1994) Somatic hybridization by microfusion of protoplasts. In: Bajaj YPS (ed) Somatic hybridization in crop improvement I. Biotechnology in agriculture and forestry, vol 27. Springer, Berlin Heidelberg New York, pp 44–56

Spangenberg G, Vallés MP, Wang ZY, Montavon P, Nagel J, Potrykus I (1994) Asymmetric somatic hybridization between tall fescue (*Festuca arundinacea* Schreb.) and irradiated Italian ryegrass (*Lolium multiflorum* Lam.) protoplasts. Theor Appl Genet 88: 509–519

Spangenberg G, Wang ZY, Legris G, Montavon P, Takamizo T, Pérez-Vicente R, Vallés MP, Nagel J, Potrykus I (1995) Intergeneric symmetric and asymmetric somatic hybridization in *Festuca* and *Lolium*. Euphytica 85: 235–245

Tabaeizadeh Z, Ferl RJ, Vasil IK (1986) Somatic hybridization in the Gramineae: *Saccharum officinarum* L. (sugarcane) + *Pennisetum americanum* (L.) K. Schum. (pearl millet). Proc Natl Acad Sci USA 83: 5616–5619

Takamizo T, Spangenberg G (1994) Somatic hybridization in *Festuca* and *Lolium*. In: Bajaj (ed) Somatic hybridization in crop improvement I. Biotechnology in agriculture and forestry, vol 27. Springer, Berlin Heidelberg New York, pp 112–131

Takamizo T, Spangenberg G, Suginobu K, Potrykus I (1991) Intergeneric somatic hybridization in Gramineae: somatic hybrid plants between tall fescue (*Festuca arundinacea* Schreb.) and Italian ryegrass (*Lolium multiflorum* Lam.). Mol Gen Genet 231: 1–6

Terada R, Kyozuka J, Nishibayashi S, Shimamoto K (1987) Plantlet regeneration from somatic hybrids of rice (*Oryza sativa* L.) and barnyard grass (*Echinochloa oryzicola* Vasing). Mol Gen Genet 210: 39–43

Van der Maas HM, Zaal MACM, de Jong ER, Van Went JL, Krens FA (1993a) Optimization of isolation and storage of sperm cells from pollen of perennial ryegrass (*Lolium perenne* L.). Sex Plant Reprod 6: 64–70

Van der Maas HM, Zaal MACM, de Jong ER, Krens FA, Van Went JL (1993b) Isolation of viable egg cells of *Lolium perenne* L. Protoplasma 173: 86–89

Van der Maas HM, de Jong ER, Van Aelst AC, Verhoeven HA, Krens FA, Van Went JL (1994) Cytological characterization of isolated sperm cells of perennial ryegrass (*Lolium perenne* L.). Protoplasma 178: 48–56

Van Ark HF, Hall RD, Creemers-Molenaar J, Krens FA (1992) High yields of cytoplasts from protoplasts of *Lolium perenne* and *Beta vulgaris* using gradient centrifugation. Plant Cell Tissue Organ Cult 31: 223–232

Vasil IK (1987) Developing cell and tissue culture systems for the improvement of cereals and grass crops. J Plant Physiol 128: 193–218

Vasil V, Ferl RJ, Vasil IK (1988) Somatic hybridization in the Gramineae: *Triticum monococcum* L. (einkorn) + *Pennisetum americanum* (L.) K. Schum. (pearl millet). J Plant Physiol 132: 160–163

Yang ZQ, Shikanai T, Mori K, Yamada Y (1989) Plant regeneration from cytoplasmic hybrids of rice (*Oryza sativa* L.). Theor Appl Genet 77: 305–310

Chapter 7
Transgenic Plants from Protoplasts

7.1 Introduction

Direct gene transfer to protoplasts is one of several methods which have been developed for the generation of transgenic plants and for transient gene expression studies (Potrykus 1990, 1991, 1995; Saul and Potrykus 1990). The method is based on the efficient uptake of plasmid DNA from the surrounding medium into protoplasts which is promoted either by chemical treatment, e.g. with polyethylene glycol (PEG; Paszkowski et al. 1984; Negrutiu et al. 1987), or by the application of electric pulses (electroporation; Fromm et al. 1986), or by a combination of both (Shillito et al. 1986). In those cases where an appropriate protoplast-to-plant regeneration system is available, a large number of transformed clones can be obtained, and often these clones can be regenerated to fully fertile transgenic plants (Davey et al. 1989; Roest and Gilissen 1989; Potrykus 1995).

The basic procedure involves the preparation of protoplasts by enzymatic digestion, the addition of DNA to the protoplast suspension, the uptake of DNA stimulated by one of the various treatments indicated above, the selection – usually applied at some point in the regeneration process from the treated protoplast to callus – for the expression of a transformed gene conferring a selectable phenotype, and finally, the stimulation of morphogenesis on developing calli leading to plant regeneration (Saul and Potrykus 1990; Potrykus 1995). After optimization of parameters of the protocol, high efficiencies of integrative transformation can be achieved (Negrutiu et al. 1987). In most cases, transgenic plants from protoplasts show stable inheritance of the foreign gene predominantly as a single locus (Potrykus et al. 1985a). Based on direct gene transfer to protoplasts, co-transformation of two plasmids bearing a selectable and a non-selectable gene is possible at a high frequency (Schocher et al. 1986). In addition, the method allows for gene targeting based on homologous recombination, though at a very low frequency (Paszkowski et al. 1988).

Reports on the transient expression of foreign genes in grass protoplasts, the production of true transgenic forage and turf grass plants by direct gene transfer to protoplasts, and the surrogate transformation of grasses with protoplast-derived transformed grass endophytes are reviewed below with special consideration of studies on fescues and ryegrasses (Table 7.1).

Table 7.1. Direct gene transfer to protoplasts for transformation of forage and turf grasses

Plant species	Transgene	Outcome	References
Festuca arundinacea (tall fescue)	*hph*, *bar*	Transgenic plants	Wang et al. (1992)
	hph, *gusA*	Putative transgenic plants	Ha et al. (1992)
	gusA	Transfected protoplasts	Penmetsa & Ha (1994)
	hph, *bar*	Transgenic plants	Spangenberg et al. (1995)
	hph	Transgenic plants	Dalton et al. (1995)
	gusA	Transfected protoplasts	Ha et al. (1995)
	gusA	Transformed calli	Kuai and Morris (1996)
	hph	Transformed endophyte and surrogate transformed plants	Tsai et al. (1992)
Festuca pratensis (meadow fescue)	*bar*	Transgenic plantlets	Spangenberg et al. (1995)
Festuca rubra (red fescue)	*bar*	Transgenic plants	Spangenberg et al. (1994)
Lolium multiflorum (Italian ryegrass)	*npt2*	Transformed calli	Potrykus et al. (1985b)
	npt2, *gusA*	Putative transgenic plants	Wang et al. (1997)
Lolium perenne (perennial ryegrass)	*hph*, *gusA*	Transformed endophyte and surrogate transformed plants	Murray et al. (1992)
	npt2, *gusA*	Putative transgenic plants	Wang et al. (1997)
Agrostis alba (redtop)	*gusA*	Transfected protoplasts	Asano et al. (1991)
	npt2	Transgenic plants	Asano & Ugaki (1994)
Dactylis glomerata (cocksfoot)	*hph*	Transgenic plants	Horn et al. (1988)
Panicum maximum (guinea grass)	*cat*	Transfected protoplasts	Hauptmann et al. (1987)
	cat	Transfected protoplasts	Vasil et al. (1988)
	dhfr	Transformed calli	Hauptmann et al. (1988)
Pennisetum purpureum (napier grass)	*cat*	Transfected protoplasts	Hauptmann et al. (1987)

Transgenes: *bar*, *pat*: phosphinothricin acetyltransferase; *cat*: chloramphenicol acetyltransferase; *dhfr*: dihydrofolate reductase; *gusA*: β-glucuronidase; *hph*: hygromycin phosphotransferase; *npt2*: neomycin phosphotransferase II

7.2 Direct Gene Transfer to Protoplasts in *Festuca*

A key prerequisite for the generation of transgenic *Festuca* plants by direct gene transfer is the availability of a reproducible and efficient plant regeneration system from embryogenic cell suspension-derived protoplasts. The establishment of such protoplast-to-plant regeneration systems has been reported for different fescues: *F. arundinacea* (Dalton 1988a, b; Takamizo et al. 1990), *F. pratensis* (Wang et al. 1993) and *F. rubra* (Spangenberg et al. 1994; Chap. 5), thus opening up opportunities for the recovery of transgenic fescue plants from protoplasts. Furthermore, functional expression vectors bearing selectable marker genes and tight selection schemes after transfection of protoplasts are needed.

In *Festuca*, transgenic soil-grown plants were first obtained for different forage-type cultivars of tall fescue by using PEG-mediated transformation of cell suspension-derived protoplasts (Wang et al. 1992). A chimeric hygromycin phosphotransferase (HPH) gene (*hph*) and two chimeric phosphinothricin acetyltransferase (PAT) genes (*bar* and *pat*) were used as selectable markers in this study. To gain information as to whether or not plasmid DNA carrying these chimeric genes was introduced into the PEG-treated tall fescue protoplasts and, if so, whether it was functional, transient expression was assayed after transfection (Wang et al. 1992). HPH activity well above the background from the control non-transfected protoplasts was detected in protein extracts from protoplasts treated with the chimeric *hph* gene driven by the cauliflower mosaic virus (CaMV) 35S promoter. Similarly, significant PAT activity was found after PEG-mediated transfection of tall fescue protoplasts when the CaMV 35S driven chimeric *bar* and synthetic *pat* genes were used (Wang et al. 1992; Spangenberg et al. 1995).

A further prerequisite for the recovery of transgenic plants is the availability of a tight selection scheme to enrich those protoplasts and derived callus clones which stably express the chimeric selectable marker genes. The stable integration and expression of a chimeric *hph* gene results in the ability to phosphorylate hygromycin, thus rendering the transformed plant cells resistant to concentrations of this antibiotic which are lethal to untransformed cells. Dose response experiments with agarose-embedded protoplasts of *F. arundinacea* in bead-type culture in which hygromycin in the concentration range of 25–400 mg/l was used, showed a high level of natural tolerance to this antibiotic (Wang et al. 1992). Levels of up to 100 mg/l hygromycin still allowed growth of colonies, whereas 200 mg/l hygromycin during bead-type culture of transfected tall fescue protoplasts followed by plate selection of resistant calli onto proliferation medium with 200 mg/l hygromycin provided tight selection (Spangenberg et al. 1995). The expression of *bar* and *pat* chimeric genes in transfected protoplasts results in the acetylation of phosphinothricin and thus renders plant cells resistant to this herbicide. Dose response experiments to determine lethal concentrations of phosphinothricin for tall fescue protoplasts were performed with three different media based on standard AA medium (Müller and Grafe 1978), a modified AA medium with glutamine replaced by asparagine (Wang et al. 1992), and the General medium (Chen 1986) free of amino acids. Wang et al. (1992) found for non-selected untransformed tall fescue protoplasts that the General medium gave a lower plating efficiency than standard AA medium. Plating efficiencies of tall fescue protoplasts in AA medium containing asparagine instead of glutamine were even lower, on average 30% of that for standard AA medium. Lethal phosphinothricin concentrations for tall fescue protoplasts were 25 mg/l in the case of the General medium and 50 mg/l in the case of standard AA-based medium used (Spangenberg et al. 1995). A selection scheme for tall fescue protoplasts transfected with chimeric *bar* or *pat* genes in which a supralethal dose of 100 mg/l phosphinothricin was used during bead-type culture was found to prevent growth of non-transformed calli (Wang et al. 1992; Spangenberg et al. 1995). After direct gene transfer, the tall fescue protoplasts (Fig. 7.1A) were plated on agarose-solidified AA medium and underwent multiple divisions within 6–10 days after transfer of the agarose beads to liquid medium with nurse cells. Cell divisions continued and the nurse cells could be removed about 1 week later. The plating efficiency of tall fescue protoplasts (number of visible colonies/ number of plated protoplasts) after PEG treatment was generally lower (0.04%) than that of untreated protoplasts (0.06%; Wang et al. 1992; Spangenberg et al. 1995).

After 3–4 weeks in selection (using supralethal doses of hygromycin or phosphinothricin), putative transformed colonies of about 2–5 mm in diameter or complete agarose beads containing hygromycin-resistant colonies from *F. arundinacea* (Fig. 7.1B) were transferred onto solidified medium containing 200 mg/l hygromycin for transformations with the chimeric *hph* gene but no phosphinothricin in the case of transformations with the chimeric *bar* or *pat* genes (Fig. 7.1C). Overall transformation frequencies (number of resistant colonies/number of treated protoplasts) in the range of 10^{-5} to 10^{-6} were found (Wang et al. 1992; Spangenberg et al. 1995). Resistant colonies continued to proliferate on the medium without selection until regenerating embryogenic callus clones were established (Fig. 7.1D). These tall fescue resistant calli regenerated multiple green shoots and in vitro-grown plantlets at frequencies comparable to non-transformed tall fescue calli (Fig. 7.1D, E). On average,

Fig. 7.1A–G. Recovery of transgenic plants from protoplasts in *Festuca arundinacea*. **A** Protoplasts isolated from embryogenic cell suspension of tall fescue immediately after transformation with plasmid pGL2. **B** Agarose bead with resistant colonies growing at a hygromycin concentration of 200 mg/l 5 weeks after protoplast plating. **C** Hygromycin-resistant calli on proliferation medium 2 months after transformation. **D** Regeneration of shoots from hygromycin-resistant callus clones. **E** Regenerated hygromycin-resistant plantlets derived from independent selected calli. **F** Independent transgenic mature plants of tall fescue potted in soil and growing under containment greenhouse conditions. **G** Flowering transgenic hygromycin resistant tall fescue plant. (Wang et al. 1992)

A
B
C
D
E
F
G

3–5 months after direct gene transfer to protoplasts, rooted tall fescue transgenic plants were transferred to soil and grown until maturity (Fig. 7.1F, G; Wang et al. 1992; Spangenberg et al. 1995). A detailed molecular characterization of these first transgenic forage-type tall fescue plants from protoplasts was performed (Wang et al. 1992). Representative results from these analyses are shown for tall fescue transformants obtained from direct gene transfer experiments with plasmid pGL2 bearing a chimeric *hph* gene (Fig. 7.2). Initial screening by polymerase chain reaction (PCR) using primers designed to amplify an internal 375-bp *hph* fragment revealed the presence of amplification products of the expected size for independent hygromycin resistant clones and their absence in the untransformed tall fescue negative control (Fig. 7.2A, B). Southern hybridization analysis using total cellular DNA was performed with undigested and digested DNA samples for independent hygromycin-resistant tall fescue plants (Fig. 7.2C, D). This analysis confirmed the integration of multiple full length transgene copies into the genome of the transgenic plants as well as the occurrence of transgene rearrangements (Wang et al. 1992). In situ hybridization experiments on metaphase spreads indicated that multiple transgene copies were integrated into one chromosomal location (Fig. 7.2E). The integration of functional copies was proven by the respective enzyme in all transgenic tall fescue plants tested (Fig. 7.2F).

Ha et al. (1992) have reported the generation of putative transgenic calli and plants after electroporation of protoplasts for one turf-type cultivar of *F. arundinacea*. Chimeric *hph* and β-glucuronidase (GUS) genes (*gusA*) driven

→

Fig. 7.2A–F. Molecular analysis of transgenic plants of *Festuca arundinacea* obtained from protoplasts. **A** Map of pGL2 containing a chimeric *hph* gene used for direct gene transfer to tall fescue protoplasts. **B** Ethidium bromide-stained agarose gel of PCR amplification products of hygromycin-resistant tall fescue in vitro-growing plantlets. *Lanes 1–8* Eight independent hygromycin-resistant regenerants; *9* non-transformed tall fescue negative control; *10* positive control pGL2. **C** Southern blot hybridization with total genomic DNA (not digested) from hygromycin-resistant tall fescue plants hybridized with *hph* probe. *Lanes 1–8* Eight independent hygromycin-resistant regenerants; *9* non-transformed tall fescue negative control. DNA samples are from the same hygromycin-resistant plants as in **B**. **D** Southern blot hybridization analysis of undigested (*lane 1*) and digested (*lane 2*: *Eco*RI, *lane 3*: *Hind*III, *lane 4*: *Bam*HI) DNA from a transgenic plant. DNA from untransformed tall fescue control plant was digested with *Eco*RI (*lane 5*) and *Bam*HI (*lane 6*). *Bam*HI-digested pGL2 (5 pg and 10 pg) was loaded onto *lanes 7* and *8*. Integration of multiple and rearranged copies is shown. **E** In situ hybridization of *hph* sequence to metaphase chromosomes of transgenic tall fescue. Digoxigenin-labeled *hph* sequence (*Bam*HI fragment from pGL2) was used as probe. In situ hybridization of labeled probe to one metaphase chromosome from the transgenic plant shown in **D** is marked with an *arrow*. The extent of hybridization of the digoxigenin-labeled probe was revealed by 4-nitroblue tetrazoliumchloride reduction catalyzed by alkaline phosphatase coupled to anti-digoxigenin antibodies, with *bar* representing 10 μm. **F** Expression of hygromycin phosphotransferase in transgenic plants derived from protoplasts transformed with plasmid pGL2. *Lanes 1, 2* Untransformed tall fescue plant as negative control; *3–10* four independent transgenic tall fescue plants; *11, 12* transformed hygromycin-resistant tobacco plant as positive control. Assays were performed in the presence (*odd-numbered lanes*) or absence (*even-numbered lanes*) of hygromycin. (Wang et al. 1992)

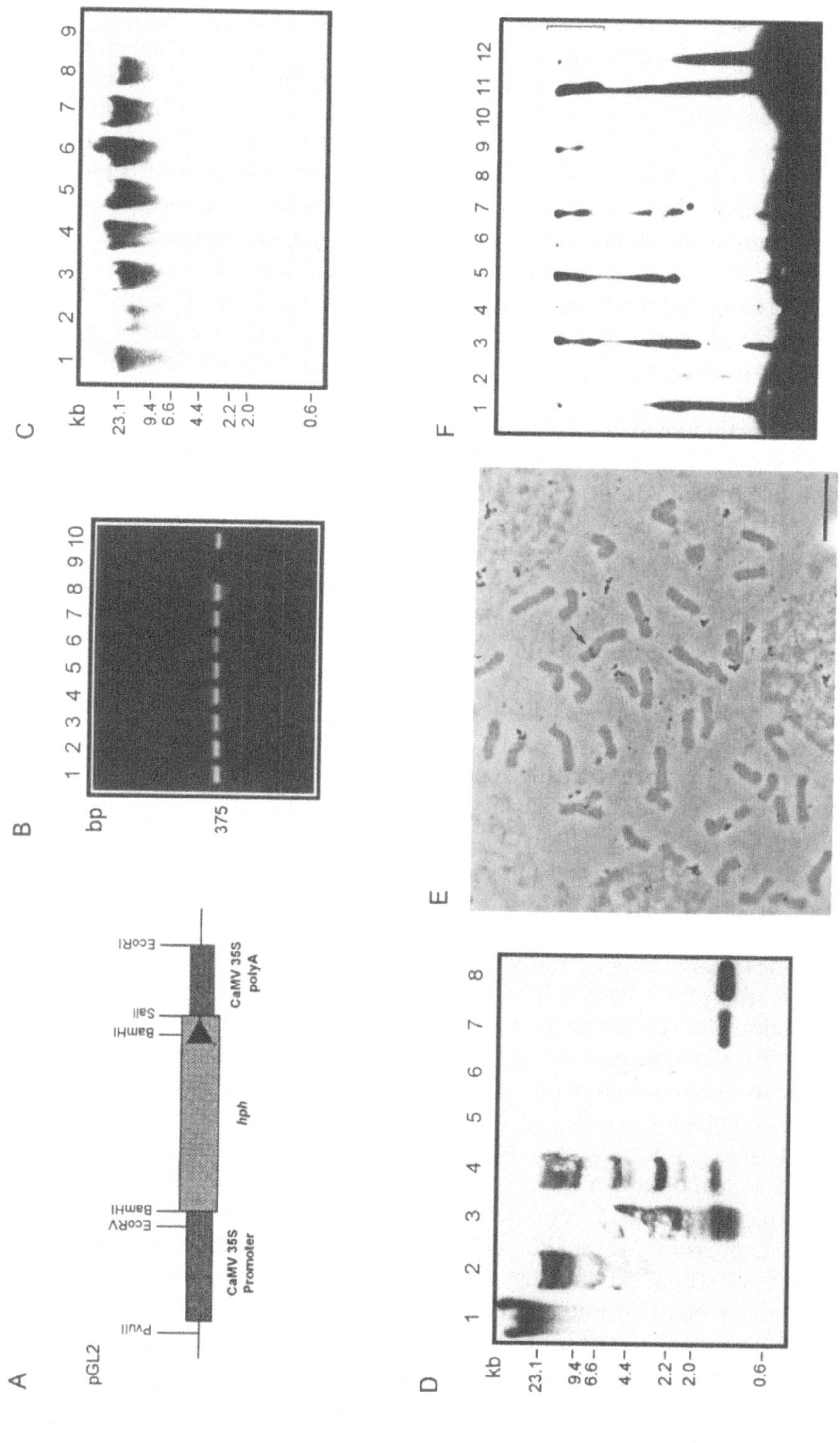
A
pGL2
PvuII
EcoRV
BamHI
BamHI
SalI
EcoRI
CaMV 35S Promoter
hph
CaMV 35S polyA
B
bp
375
1 2 3 4 5 6 7 8 9 10
C
kb
23.1
9.4
6.6
4.4
2.2
2.0
0.6
1 2 3 4 5 6 7 8 9
D
kb
23.1
9.4
6.6
4.4
2.2
2.0
0.6
1 2 3 4 5 6 7 8
E
F
1 2 3 4 5 6 7 8 9 10 11 12

by CaMV 35S promoters were used as selectable marker and reporter genes, respectively. The transfected tall fescue protoplasts were selected in liquid culture medium containing 150 mg/l hygromycin and by subsequent plate selection with 200 mg/l of the selective agent. The overall transformation frequency was $3–9 \times 10^{-6}$ (Ha et al. 1992). Histochemical GUS activity was found in most of the hygromycin-resistant colonies developed from transfected tall fescue protoplasts. Expression of the *gusA* reporter gene was also detected in leaf tissue from plants regenerated from hygromycin-resistant calli. Southern hybridization analysis was performed for PCR amplification products only, thus providing circumstantial evidence supporting the transgenic nature of the plants recovered (Ha et al. 1992). However, conclusive experimental data unequivocally demonstrating integrative transformation are still lacking for this plant material.

Selection schemes for the recovery of transgenic tall fescue plants from protoplasts PEG-transfected with a chimeric *hph* gene have been partially optimized by Dalton et al. (1995). The effect of selection pressure on the overall transformation frequency and transgene copy number in the transgenic *F. arundinacea* plants recovered was evaluated. Discontinuous and continuous selection schemes with 50–200 mg/l hygromycin were tested. Significant numbers of escapes were obtained when the discontinuous selection scheme was used. In contrast, continuous selection at relatively low levels (50 mg/l) of hygromycin allowed for recovery of a relatively high number of transgenic plants without escapes (Dalton et al. 1995). The overall transformation frequencies varied between 10^{-5} and 10^{-6}, thus being well within the range of previous reports for this species (Ha et al. 1992; Wang et al. 1992). The transgene copy numbers of a large number of transgenic tall fescue plants from protoplasts obtained from the different selection regimes were determined (Dalton et al. 1995). In total, out of the 83 plants analyzed, 75% contained two or fewer copies and 10% contained five copies or more. Single-copy transgenics were obtained in 33% of the cases. Severe continuous selection regimes at high concentrations of hygromycin favoured selection of transgenic tall fescue plants having integrated multiple transgene copies. Discontinuous selection allowed for recovery of a high proportion (87%) of plants containing two or fewer transgene copies. Similar results were obtained when continuous selection with 50 mg/l hygromycin was used. Transgene copy number had no effect on mitotic stability of transgene expression when soil-grown transgenic tall fescue plants were re-tested after 1 year (Dalton et al. 1995).

Experimental conditions for the transfection of tall fescue protoplasts by electroporation have been partially optimized on the basis of transient *gusA* gene expression assays (Penmetsa and Ha 1994). Maximum GUS activity was measured when 750 V/cm field strength and 400 μF capacitance were used. The effect of DNA concentration on GUS activity after transfection of tall fescue protoplasts was tested in the 8- to 200-μg/ml range. Increases in the amount of plasmid DNA to 80 μg/ml led to increases in GUS activity (Penmetsa and Ha 1994). The effect of protoplast density on transient GUS expression was tested in the range of 0.5–6 $\times$ 10^6 protoplasts/ml. Measured GUS activity was found to increase with increases in protoplast density during transfection of up to 2 $\times$ 10^6 protoplasts/ml. An effect on GUS activity was found when tall fescue protoplasts isolated from suspension cultures at different time points in their growth curves were transfected. Highest activities were found when the protoplasts were derived from rapidly growing (3–5 days after subculture) suspension cells. A gradual decline in GUS activity was detected when slower growing (5–7 days after subculture) and stationary (7–10 days after subculture) cultures were used as a source of protoplasts for transfections and transient gene expression assays (Penmetsa and Ha 1994). Five chimeric *gusA* genes driven by different 5′ regulatory sequences [CaMV 35S promoter alone, CaMV 35S promoter plus alcohol dehydrogenase 1 (*adh1*) intron 1, CaMV 35S promoter plus *adh1* intron 6, rice actin 1 (*act1*) promoter plus *act1* intron 1, and maize *adh1* promoter plus *adh1* intron 1] were evaluated using this protocol for electroporative transfection of tall fescue protoplasts (Ha et al. 1995). Highest GUS expression activities were detected when the rice *act1* 5′ regulatory sequences including *act1* intron 1 were used.

Transgenic *F. rubra* plants from protoplasts have been obtained through PEG-mediated transformation using chimeric *bar* genes driven by rice *act1* 5′ regulatory sequences (Spangenberg et al. 1994). The transfected red fescue protoplasts were selected in bead-type culture containing 50 mg/l phosphinothricin for 3–4 weeks prior to transfer of the phosphinothricin-resistant colonies onto selective solidified medium for further proliferation. Overall transformation frequencies were in the range of 0.5–2 $\times$ 10^{-6}. Resistant callus clones transferred onto regeneration medium differentiated multiple plantlets through somatic embryogenesis in 50% of the cases (Spangenberg et al. 1994). Five to 6 months after direct gene transfer to protoplasts, in vitro-rooted plantlets could be transferred to soil and grown further under greenhouse conditions. The transgenic nature of the red fescue plants obtained from protoplasts was confirmed by Southern hybridization analysis in which digested and non-digested DNA samples were used. Hybridization signals corresponding to high molecular weight bands (for the non-digested DNA samples) and to the *bar* gene expression cassette (for the accordingly digested DNA samples) were detected. In addition to bands of the expected size, a complex pattern – indicating multiple copy integrations and rearrangements of the transgene – was observed. Adult transgenic red fescue plants were shown to survive spraying with the phosphinothricin-based herbicide Basta using doses supralethal for the untransformed control plant in bead-type culture (Spangenberg et al. 1994).

Direct gene transfer to protoplasts isolated from single genotype-derived embryogenic cell suspensions has also been reported for *F. pratensis* (Spangenberg et al. 1995). The meadow fescue protoplasts were subjected to PEG-mediated transformation using a chimeric *bar* gene driven by the CaMV 35S promoter. Selection of transfected protoplasts at a phosphinothricin concentration of 100 mg/l was found to be tight. Stable transformation frequencies were between 10^{-5} and 10^{-6}. In vitro-grown plantlets were established from the phosphinothricin-resistant calli. All phosphinothricin-resistant plantlets were found to be PCR-positive when subjected to a screening in which primers designed to amplify an internal 310-bp *bar* fragment were used (Spangenberg et al. 1995). The true *bar* identitiy of the amplification products obtained from PCR reactions with DNA samples from the phosphinothricin-resistant meadow fescue regenerants was confirmed by Southern blot hybridization analysis. Further evidence supporting the transgenic nature of the phosphinothricin-resistant plantlets of meadow fescue recovered from protoplasts was obtained from genomic Southern hybridization analysis in which undigested samples of total cellular DNA were used. This analysis revealed hybridization signals corresponding to high molecular weight bands, thus indicating the integration of the *bar* gene into the plant genome of the regenerants analyzed (Spangenberg et al. 1995).

The experimental basis for the surrogate transformation of *F. arundinacea* by infection of wild type seedlings with a transformed endophyte has been established (Tsai et al. 1992). Protoplasts prepared from mycelia of *Neotyphodium coenophialum* were electroporated with chimeric *hph* genes driven by 5′ regulatory sequences of the *Epichloe typhina* β-tubulin gene (*tub2*) and selected at a hygromycin concentration of 48 mg/l. Transformation frequencies (number of resistant fungal colonies/number of electroporated protoplasts) of fungal protoplasts varied between 10^{-5} and 10^{-7}. Different parameters affecting transformation frequencies were evaluated. Particularly the concentration of plasmid DNA and the fungal protoplast density during electroporation were found to be important. Corresponding optimal values were determined, i.e. 6 μg of plasmid DNA and 2×10^9 protoplasts/ml. The generation of stably transformed *N. coenophialum* mycelia from electroporated protoplasts was demonstrated by Southern hybridization analysis. The transformants analyzed showed the integration of single and also multiple or rearranged copies of the chimeric *hph* gene used. The *N. coenophialum* transformants were inoculated into endophyte-free *F. arundinacea* seedlings. Infections of transgenic endophytes in otherwise wild-type tall fescue plants succeeded in 9–27% of the cases. No information on transgene expression in the transformed mycelium in vivo within the *F. arundinacea* surrogate-transformed host plants was provided (Tsai et al. 1992).

7.3 Direct Gene Transfer to Protoplasts in *Lolium*

Direct gene transfer to protoplasts for a graminaceous monocot was first reported in *Lolium* (Potrykus et al. 1985b). Confirmed stably transformed callus clones were obtained for *L. multiflorum* using a chimeric neomycin phosphotransferase (NPT2) gene (*npt2*) driven by the CaMV 19S promoter through PEG treatment of non-morphogenic protoplasts. Stable expression of the *npt2* gene in plant cells render them resistant to the antibiotics kanamycin and geneticin (G418). Transfected Italian ryegrass protoplasts were selected in liquid medium containing 25 mg/l G418 for 4 weeks and the resistant colonies then transferred onto solidified proliferation medium still containing the same level of G418 as selective agent. Selection with kanamycin levels of up to 800 mg/l proved to be ineffective (Potrykus et al. 1985b). A transformation frequency of 2.5×10^{-4} was estimated. Most of the G418-resistant calli proved to be transformed, as confirmed by Southern hybridization analysis performed with non-digested and digested DNA samples. Evidence for the integration of multiple copies of the chimeric *npt2* gene was obtained. The functionality of some of the integrated *npt2* transgene copies was demonstrated by corresponding NPT2 enzyme assays (Potrykus et al. 1985b).

In spite of significant progress made in the regeneration of fertile plants from protoplasts prepared from embryogenic suspension cells of different *Lolium* species (Table 5.1), so far only circumstantial evidences on the recovery of first transgenic ryegrass plants from protoplasts have been provided (Wang et al. 1997). A thorough molecular characterization of the primary putative transgenic *L. perenne* and *L. multiflorum* plants – obtained after direct gene transfer to protoplasts with chimeric *npt2* and *gusA* genes – and of the corresponding progenies is required to unequivocally confirm their true transgenic nature.

Surrogate transformation of *L. perenne* involving transformed *Neotyphodium* endophytes produced by direct gene transfer to the fungal protoplasts has been described (Murray et al. 1992). Both linearized and circular plasmids bearing chimeric *hph* and *gusA* genes driven by *Aspergillus nidulans* glyceraldehyde-3-phosphate dehydrogenase (*gpd*) promoter sequences were co-transferred into fungal endophyte protoplasts by PEG-mediated transformation. After plate selection with 200 mg/l hygromycin, resistant fungal colonies were obtained (Fig. 7.3A). Transformation frequencies were estimated to be in the range of $3–4 \times 10^{-4}$. The stability of the hygromycin resistance phenotype upon further culture was 25 and 80% of the initially recovered transformants when these were derived from protoplast transfections using circular and linearized plasmids, respectively. Expression of the reporter *gusA* gene was observed in 80% of the hygromycin-resistant transformants (Fig. 7.3A). Integration of the transgenes was demonstrated by Southern hybridization analysis. Both single and tandemly repeated copies of the plasmid were integrated in the genome of different *Neotyphodium* transformants (Murray et al. 1992). Several *hph* and *gusA* expressing *Neotyphodium* transformants were inoculated into endophyte-free perennial ryegrass seedlings. Expression of the *gusA* gene in the transgenic endophyte growing intercellularly in vivo in the surrogate-transformed *L. perenne* plants was demonstrated by histochemical GUS assays (Fig. 7.3B). Southern hybridization analysis of transgenic fungal isolates from the leaf sheath of surrogate-transformed perennial ryegrass plants revealed patterns identical to that observed in the fungus when used as an inoculum (Murray et al. 1992).

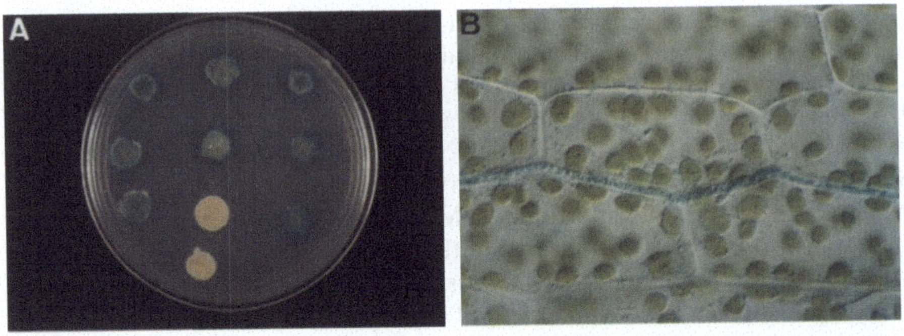

Fig. 7.3A,B. Histochemical staining for β-glucuronidase (GUS) activity in transformants of *Neotyphodium* (formerly *Acremonium*) in free living culture and in planta in the surrogate transformed *Lolium perenne* host. **A** Fungal cultures plated on medium containing 5-bromo-4-chloro-3-indolyl β-D-glucuronic acid and photographed after 10 days. Nine blue-stained colonies from GUS-positive *Neotyphodium* co-transformants obtained from protoplasts expressing *hph* and *gusA* genes are shown. Two white colonies are from the negative control, the untransformed *Neotyphodium* strain 187BB, and a transformant expressing only the *hph* gene, rescpectively. **B** Histochemical detection of GUS activity in the transformed *Neotyphodium* endophytic mycelium growing intercellularly within the leaf sheath of the surrogate-transformed perennial ryegrass host plant. (Murray et al. 1992; Courtesy of DB Scott, Molecular Genetics Unit, Department of Microbiology and Genetics, Massey University, Palmerston North, New Zealand)

7.4 Direct Gene Transfer to Protoplasts in Other Grasses

Transient gene expression studies using transfected protoplasts and the recovery of transgenic plants from protoplasts have been reported for different grass species outside the *Festuca-Lolium* complex, namely *Agrostis alba* (Asano et al. 1991; Asano and Ugaki 1994), *Dactylis glomerata* (Horn et al. 1988), *Panicum maximum* (Hauptmann et al. 1987, 1988; Vasil et al. 1988), and *Pennisetum purpureum* and the trispecific hybrid *P. purpureum* × *P. americanum* × *P. squamulatum* (Hauptmann et al. 1987; Vasil et al. 1988; Table 7.1).

Transient GUS expression was assessed after transfection of *A. alba* protoplasts by electroporation with a chimeric CaMV 35S-driven *gusA* gene on a standard pUC-derived vector or on a wheat dwarf virus (WDV)-derived plant replicative vector (Asano et al. 1991). Electroporation-mediated DNA delivery was partially optimized under the conditions for embryogenic cell suspension-derived redtop protoplasts. Highest GUS activity was found when 750 V/cm field strength was applied for 38 ms. Expression levels of the reporter gene were higher when the plant replicative vector was used (Asano et al. 1991). Based on this protocol for electroporation-mediated direct gene transfer to protoplasts, transgenic redtop plants have been recovered (Asano and Ugaki 1994). A chimeric *npt2* gene under control of an enhanced CaMV 35S promoter and selection of transfected protoplasts at a G418 concentration of 20 mg/l were used in this study. Over 50 antibiotic resistant colonies were obtained, and overall transformation frequencies were estimated to be in the range of $1–3 \times 10^{-6}$. Southern hybridization analysis of digested DNA samples from six plants regenerated from G418-resistant calli provided evidence for the integration of one or more transgene copies. Furthermore, NPT2 enzyme activity was detected in protein extracts from the transgenic plants recovered from protoplasts (Asano and Ugaki 1994).

The first putative transgenic plants from protoplasts from a grass species were reported by Horn et al. (1988) for *Dactylis glomerata*. Protoplasts newly isolated from embryogenic suspension cultures were subjected to electroporation or PEG treatment in the presence of plasmid DNA bearing a chimeric *hph* gene driven by the CaMV 35S promoter. Selection of treated protoplasts in bead-type culture containing 20 mg/l hygromycin led to the recovery of 95 resistant cell suspension lines. Although dose response experiments performed with untransformed orchardgrass protoplasts indicated that a hygromycin level of 20 mg/l was reasonably safe for selection, the molecular analysis of putative transformed calli revealed up to 60% false positives. Both transfection protocols tested gave rise to resistant lines, and putative transformed plants were obtained and established in soil. Southern hybridization analysis revealed *hph* homologous sequences for digested DNA samples prepared from several resistant callus lines and one putative transgenic plant. Both simple and complex multicopy rearranged patterns with copy numbers varying from 1 to more than 5 were found. However, no conclusive data for the integration of the foreign gene into the genome of the putative transformed callus clones and plant was provided (Horn et al. 1988).

Studies on transient expression of chloramphenicol acetyltransferase (CAT) chimeric genes (*cat*) after direct DNA uptake of protoplasts mediated by PEG treatment or electroporation have been described for *Panicum maximum*, *Pennisetum purpureum* and the double-cross trispecific hybrid *P. purpureum* × *P. americanum* × *P. squamulatum* (Hauptmann et al. 1987, 1988; Vasil et al. 1988). For transfected *P. maximum* protoplasts, *cat* gene expression was found to be slightly higher in electroporated protoplasts in comparison to PEG-mediated DNA delivery (Vasil et al. 1988). However, PEG-transfected guinea grass protoplasts showed higher plating efficiencies. Voltage and capacitance of the electroporative treatments resulting in the highest levels of CAT activity were determined for transfected protoplasts isolated from non-morphogenic cell suspensions of *P. purpureum* and the double-cross trispecific *Pennisetum* hybrid (Hauptmann et al. 1987). In a subsequent study, the same authors evaluated the potential of *npt2*, *hph* and the dihydrofolate reductase gene (*dhfr*), which confers resistance to methotrexate, as selectable markers for stable transformation of *Pennisetum* and *Panicum* protoplasts (Hauptmann et al. 1988). Resistant calli were obtained only for electroporated *P. maximum* protoplasts selected with 0.5 mg/l methotrexate upon direct gene transfer with a chimeric *dhfr* gene. Transformation frequencies varied between 10^{-5} and 2×10^{-6}. Southern hybridization analysis performed for different *P. maximum* methotrexate-resistant clones revealed the integration of one to ten *dhfr* transgene copies per haploid genome. No transgenic plant was recovered from these transformed calli (Hauptmann et al. 1988).

7.5 Summary and Conclusions

The experimental basis for the generation of transgenic plants from protoplasts, namely the establishment of: (1) efficient protoplast-to-plant regeneration systems; (2) conditions allowing for tight in vitro selection of transformed clones; and (3) appropriate plant expression vectors tested for their functionality in transient expression studies with target protoplasts, is now in place for representative fescues and ryegrasses.

Proven transgenic plants from protoplasts within the *Festuca-Lolium* complex have been produced so far in *F. arundinacea*, *F. rubra* and *F. pratensis*. Putative transgenic plants from protoplasts have been reported for *L. perenne* and *L. multiflorum*. The transgenic character of the regenerated plants was considered as proven when: (1) molecular evidence (e.g. PCR screening and Southern hybridization data) for the presence of transgene-homologous sequences in genomic DNA isolated from regenerants recovered from selected clones showing the expected resistant phenotype was provided; (2) molecular evidence (e.g. Southern and in situ hybridization data) for the integration of transgene-homologous sequences in non-digested high molecular weight plant DNA was available; and (3) data from assays (e.g. in vitro enzyme assay and herbicide spraying) confirming functional transgene products were available.

Although transgenic plants from protoplasts were recovered in different *Festuca* species, current protocols are still suboptimal. Methods allowing for the largely genotype-independent generation of fertile transgenic plants with mainly single copy transgene integration and appropriate levels of transgene expression are needed. Progress along some of these lines, such as the development of transfection protocols applicable to protoplasts prepared from single genotype-derived suspension cultures established for a wide range of cultivars and the development of improved selection regimes allowing for a tight selection of transformants with mainly single transgene copies, has been made. However, critical information, particularly, from detailed studies on the inheritance of transgenes in transgenic grass plants from protoplasts and corresponding meiotic stability of transgene expression is still not available. Until this information is provided, direct gene transfer to protoplasts will largely remain a useful tool for transient gene expression studies.

References

Asano Y, Ugaki M (1994) Transgenic plants of *Agrostis alba* obtained by electroporation-mediated direct gene transfer into protoplasts. Plant Cell Rep 13: 243–246

Asano Y, Otsuki Y, Ugaki M (1991) Electroporation-mediated and silicon carbide fiber-mediated DNA delivery in *Agrostis alba* L. (redtop). Plant Sci 79: 247–252

Chen Y (1986) Anther and pollen culture of rice. In: Hu H, Yang HY (eds) Haploids of higher plants in vitro. China Academic, Beijing; Springer, Berlin Heidelberg New York, pp 1–25

Dalton SJ (1988a) Plant regeneration from cell suspension protoplasts of *Festuca arundinacea* Schreb. (tall fescue) and *Lolium perenne* L. (perennial ryegrass). J Plant Physiol 132: 170–175

Dalton SJ (1988b) Plant regeneration from cell suspension protoplasts of *Festuca arundinacea* Schreb., *Lolium perenne* L. and *L. multiflorum* Lam. Plant Cell Tissue Organ Culture 12: 137–140

Dalton SJ, Bettany AJE, Timms E, Morris P (1995) The effect of selection pressure on transformation frequency and copy number in transgenic plants of tall fescue (*Festuca arundinacea* Schreb.). Plant Sci 108: 63–70

Davey MR, Rech EL, Mulligan BJ (1989) Direct DNA transfer to plant cells. Plant Mol Biol 13: 273–285

Fromm M, Taylor LP, Walbot V (1986) Stable transformation of maize after gene transfer by electroporation. Nature 319: 791–793

Ha SB, Wu FS, Thorne TK (1992) Transgenic turf-type tall fescue (*Festuca arundinacea* Schreb.) plants regenerated from protoplasts. Plant Cell Rep 11: 601–604

Ha SB, Penmetsa RV, Wu FS (1995) Transient expression of the beta-glucuronidase gene driven by different 5′ regulatory sequences in tall fescue protoplasts. Turfgrass Physiol Biotechnol 1: 73–81

Hauptmann RM, Ozias-Akins P, Vasil V, Tabaeizadeh Z, Rogers SG, Horsch RB, Vasil IK, Fraley RT (1987) Transient expression of electroporated DNA in monocotyledonous and dicotyledonous species. Plant Cell Rep 6: 265–270

Hauptmann RM, Vasil V, Ozias-Akins P, Tabaeizadeh Z, Rogers SG, Fraley RT, Horsch RB, Vasil IK (1988) Evaluation of selectable markers for obtaining stable transformants in the Gramineae. Plant Physiol 86: 602–606

Horn ME, Shillito RD, Conger BV, Harms CT (1988) Transgenic plants of orchardgrass (*Dactylis glomerata* L.) from protoplasts. Plant Cell Rep 7: 469–472

Kuai B, Morris P (1996) Screening for stable transformants and stability of β-glucuronidase gene expression in suspension cultured cells of tall fescue (*Festuca arundinacea*). Plant Cell Rep 15: 804–808

Müller AJ, Grafe R (1978) Isolation and characterization of cell lines of *Nicotiana tabacum* lacking nitrate reductase. Mol Gen Genet 161: 67–76

Murray FR, Latch GCM, Scott DB (1992) Surrogate transformation of perennial ryegrass, *Lolium perenne*, using genetically modified *Acremonium* endophyte. Mol Gen Genet 233: 1–9

Negrutiu I, Shillito R, Potrykus I, Biasini G, Sala F (1987) Hybrid genes in the analysis of transformation condition. Plant Mol Biol 8: 363–373

Paszkowski J, Shillito RD, Saul MW, Mandak V, Hohn T, Hohn B, Potrykus I (1984) Direct gene transfer to plants. EMBO J 3: 2717–2722

Paszkowski J, Baur M, Bogucki A, Potrykus I (1988) Gene targeting in plants. EMBO J 7: 4021–4027

Penmetsa RV, Ha SB (1994) Factors influencing transient gene expression in electroporated tall fescue protoplasts. Plant Sci 100: 171–178

Potrykus I (1990) Gene transfer to cereals: an assessment. Bio/Technology 8: 535–542

Potrykus I (1991) Gene transfer to plants: assessment of published approaches and results. Annu Rev Plant Physiol Plant Mol Biol 42: 205–225

Potrykus I (1995) Direct gene transfer to protoplasts. In: Potrykus I, Spangenberg G (eds) Gene transfer to plants. Springer, Berlin Heidelberg New York, pp 55–57

Potrykus I, Paszkowski J, Saul MW, Petruska J, Shillito RD (1985a) Molecular and general genetics of a hybrid foreign gene introduced into tobacco by direct gene transfer. Mol Gen Genet 199: 167–177

Potrykus I, Saul MW, Petruska J, Paszkowski J, Shillito RD (1985b) Direct gene transfer to cells of a graminaceous monocot. Mol Gen Genet 199: 183–188

Roest S, Gilissen LJW (1989) Plant regeneration from protoplasts: a literature review. Acta Bot Neerl 38: 1–23

Saul MW, Potrykus I (1990) Direct gene transfer to protoplasts: fate of the transferred genes. Dev Genet 11: 176–181

Schocher RJ, Shillito RD, Saul MW, Paszkowski J, Potrykus I (1986) Co-transformation of unlinked foreign genes into plants by direct gene transfer. Bio/Technology 4: 1093–1096

Shillito RD, Saul MW, Paszkowski J, Müller JM, Potrykus I (1986) High frequency direct gene transfer to plants. Bio/Technology 3: 1099–1103

Spangenberg G, Wang ZY, Nagel J, Potrykus I (1994) Protoplast culture and generation of transgenic plants in red fescue (*Festuca rubra* L.). Plant Sci 97: 83–94

Spangenberg G, Wang ZY, Vallés MP, Potrykus I (1995) Transformation in tall fescue (*Festuca arundinacea* Schreb.) and meadow fescue (*Festuca pratensis* Huds.). In: Bajaj YPS (ed) Plant protoplasts and genetic engineering VI, Biotechnology in agriculture and forestry, vol 34. Springer, Berlin Heidelberg New York, pp 183–203

Takamizo T, Suginobu KI, Ohsugi R (1990) Plant regeneration from suspension culture derived protoplasts of tall fescue (*Festuca arundinacea* Schreb.) of a single genotype. Plant Sci 72: 125–131

Tsai HF, Siegel MR, Schardl CL (1992) Transformation of *Acremonium coenophialum*, a protective fungal symbiont of the grass *Festuca arundinacea*. Curr Genet 22: 399–406

Vasil V, Hauptmann RM, Morrish FM, Vasil IK (1988) Comparative analysis of free DNA delivery and expression into protoplasts of *Panicum maximum* Jacq. (guinea grass) by electroporation and polyethylene glycol. Plant Cell Rep 7: 499–503

Wang GR, Binding H, Posselt UK (1997) Fertile transgenic plants from direct gene transfer to protoplasts of *Lolium perenne* L. and *Lolium multiflorum* Lam. J Plant Physiol 151: 83–90

Wang ZY, Takamizo T, Iglesias VA, Osusky M, Nagel J, Potrykus I, Spangenberg G (1992) Transgenic plants of tall fescue (*Festuca arundinacea* Schreb.) obtained by direct gene transfer to protoplasts. Bio/Technology 10: 691–696

Wang ZY, Vallés MP, Montavon P, Potrykus I, Spangenberg G (1993) Fertile plant regeneration from protoplasts of meadow fescue (*Festuca pratensis* Huds.). Plant Cell Rep 12: 95–100

Chapter 8
Protoplast-Independent Production of Transgenic Plants

8.1 Introduction

Problems encountered with protoplast-based methods for the generation of transgenic plants have prompted the development of alternative techniques for gene transfer in grasses. These problems relate mainly to plant regeneration from protoplasts and do not reflect specific barriers to the uptake of foreign DNA by isolated protoplasts. Examples of these difficulties are relatively low plating efficiencies and low plant regeneration frequencies from protoplasts, species and genotype dependence often observed in the regeneration process, and albinism and somaclonal variation revealed in protoplast-derived plants (Potrykus 1990). Plant regeneration from protoplasts is thus a delicate process depending upon parameters that are not under experimental control, such as wound response and genotype-dependent competence for regeneration (Vasil 1988; Potrykus 1990). Furthermore, transgenic plants recovered from protoplasts may show serious fertility constraints and undesired integration of multiple and rearranged transgene copies (Spangenberg et al. 1995a).

Protoplast-independent gene transfer techniques which have been applied to grasses are the *biolistic* method and silicon carbide whisker-mediated transformation.

The availability of methods to deliver foreign DNA directly into regenerable cells, tissues or organs overcomes limitations due to *Agrobacterium*-host specificity and reduces tissue culture-related difficulties observed with protoplasts. Biolistics or particle bombardment transformation technology involves gene delivery into intact plant cells and tissues by high-velocity microprojectiles coated with biologically active DNA (Sanford et al. 1987, 1993; Sanford 1988). Following the original observation by Klein et al. (1987) that tungsten particles could be used to transfer plasmid DNA carrying a reporter gene into onion epidermal cells, leading to transient gene expression in the bombarded cells, the method has evolved to a reliable process with significant applications in plant biology. These applications include studies on developmentally or tissue-specific regulated gene expression, gene delivery to plastids and mitochondria, and generation of transgenic and transplastomic plants (Svab et al. 1990; Christou 1992, 1993, 1996; Klein and Fitzpatrick-

Table 8.1. Protoplast-independent transformation in forage and turf grasses

Plant species	Transgene	Method	Outcome	References
Festuca arundinacea (tall fescue)	*hph*	Biolistics	Transgenic plants	Spangenberg et al. (1995a)
	gusA	Biolistics	Transient expression	Spangenberg et al. (1995a)
	hph	Whiskers	Transgenic plants	Dalton et al. (1998)
Festuca rubra (red fescue)	*hph*	Biolistics	Transgenic plants	Spangenberg et al. (1995a)
	gusA	Biolistics	Transient expression	Spangenberg et al. (1995a)
	npt2	Biolistics	Transgenic plants	Spangenberg et al. (1998)
Lolium multiflorum (Italian ryegrass)	*gusA*	Biolistics	Transient expression	Pérez-Vicente et al. (1993)
	hph, gusA	Biolistics	Transgenic plants	Ye (1997); Ye et al. (1997)
	hph, gusA	Whiskers	Transgenic plants	Dalton et al. (1998)
Lolium perenne (perennial ryegrass)	*hph*	Biolistics	Transformed calli	Hensgens et al. (1993)
	gusA	Biolistics	Transient expression	Hensgens et al. (1993)
	gusA	Biolistics	Transient expression	Pérez-Vicente et al. (1993)
	hph, gusA	Biolistics	Transformed calli	Van der Maas et al. (1994)
	hph, gusA	Biolistics	Transgenic plants	Spangenberg et al. (1995b)
	hph	Whiskers	Transgenic plant	Dalton et al. (1998)
Agrostis alba (redtop)	*gusA*	Whiskers	Transient expression	Asano et al. (1991)
Agrostis palustris (creeping bentgrass)	*gusA*	Biolistics	Transgenic plants	Zhong et al. (1993)
	bar	Biolistics	Transgenic plants	Hartman et al. (1994)
	hph, gusA	Biolistics	Transgenic plants	Xiao and Ha (1997)
Agrostis stolonifera	*hph, gusA*	Whiskers	Transgenic plant	Dalton et al. (1998)
Dactylis glomerata (cocksfoot)	*bar, gusA*	Biolistics	Transgenic plants	Denchev et al. (1997)
Pennisetum glaucum (pearl millet)	*gusA*	Biolistics	Transient expression	Taylor and Vasil (1991)

Transgenes: *bar*: phosphinothricin acetyltransferase; *gusA*: β-glucuronidase; *hph*: hygromycin phosphotransferase

McEligott 1993; Svab and Maliga 1993; Klein 1995). Different devices also based on different particle acceleration mechanisms have been developed and applied to transient and stable plant transformation studies (Sanford 1988; Batty and Evans 1992; McCabe and Christou 1993; Sautter 1993; Vain et al. 1993).

Plasmid DNA can be delivered into intact suspension-cultured cells aided by silicon carbide whiskers. Cultured suspension cells are mixed in the presence of plasmid DNA and the needle-like whiskers. The collisions between the silicon carbide whiskers and the suspension cells lead to cell penetration, DNA uptake, and stable transformation of plant cells (Kaeppler et al. 1990, 1992; Dunahay 1993). Transgenic plants can be recovered from the whiskers-treated transformed plant cells (Frame et al. 1994; Wang et al. 1995).

The following sections are devoted to studies on the application of protoplast-independent transformation methods for the analysis of gene expression in grasses and for the production of transgenic forage and turf grass plants (Table 8.1).

8.2 Protoplast-Independent Transformation in *Festuca*

Transgenic *Festuca* plants have been produced by microprojectile bombardment technology (Spangenberg et al. 1995a) and whiskers-mediated transformation (Dalton et al. 1998). Embryogenic suspension-cultured cells were used as targets for biolistic transformation. This differs from most studies on the generation of transgenic plants in cereals, where immature zygotic embryos have been used – following a short preculture phase – as targets for microprojectile bombardment, e.g. in the case of rice (Christou et al. 1991), maize (Dunder et al. 1995), wheat (Vasil et al. 1993), barley (Wan and Lemaux 1994), and sorghum (Casas et al. 1993). In the self-infertile outcrossing fescues, plants are expected to be highly heterozygous, and different seeds (embryos) within a given cultivar are expected to represent different genotypes. Furthermore, the in vitro response from plated seeds or isolated zygotic embryos shows strong genotype dependence in *Festuca*, with 3–20% of the explants producing regenerable embryogenic calli (Chap. 5). In addition, the isolation of immature zygotic embryos from small-seed grasses is more time-consuming than for large-grain cereals. These features make the use of zygotic embryos as primary targets for microballistic transformation in fescues inconvenient as compared to the cereals. Single genotype-derived embryogenic suspension cultures from responsive genotypes are ideal and reproducible sources of target cells for direct DNA delivery into the small-seed grasses.

The first transgenic fescues recovered from biolistic transformation (Fig. 8.1; Spangenberg et al. 1995a) were obtained from embryogenic cell suspensions for forage and turf-type cultivars of *F. arundinacea* (Takamizo et al. 1990; Wang et al. 1992) and *F. rubra* (Spangenberg et al. 1994). A low-cost and simple microprojectile acceleration device, the particle inflow gun (Finer et al. 1992), was chosen in this study. The delivery of DNA-coated microprojectiles to suspension-cultured cells of tall and red fescues was partially optimized using transient expression assays of a chimeric *gusA* gene construct driven by the CaMV 35S promoter. The target for microprojectile bombardment con-

sisted of approximately 250mg of fresh-weight embryogenic cell clusters plated as a monolayer on a filter paper disk (Fig. 8.1A). Different treatments of the target tall fescue and red fescue cells (e.g. pre- and post-bombardment osmotic treatments) and different DNA-particle delivery parameters (e.g. bombardment pressure and distance, baffle mesh size and baffle distance, number of bombardments, and concentration of plasmid DNA and microprojectiles) were evaluated. An approximately 100-fold increase in transient transformation events with maxima at over 2000 β-glucuronidase (GUS)-expressing foci per bombardment (Fig. 8.1B, G) was achieved. Pre-bombardment osmotic treatment of the suspension cells for 30min in liquid medium containing 30g/l sucrose and supplemented with 0.25M mannitol and 0.25M sorbitol led to a significant increase in the number of transient *gusA* expression events (Spangenberg et al. 1995a). Other microprojectile bombardment conditions found to be optimal were: (1) 3- to 4-day post-bombardment osmotic treatment of suspension cells plated on the same yet solidified medium; (2) 500-μm baffle mesh size; (3) 12-cm baffle distance; (4) 15-cm target-bombardment distance; (5) 6-bar bombardment pressure; (6) 500-μg gold particles coated with 10μg of plasmid DNA; and (7) one bombardment with a 10-μl DNA-spermidine-bound particle suspension per target (Spangenberg et al. 1995a). GUS expression was detected 3 and up to 15 days after bombardment with DNA-coated microprojectiles, indicating division and growth of the tall and red fescue cells (Fig. 8.1B, G).

Fig. 8.1A–L. Transient gene expression and stable transformation after particle inflow gun-mediated microprojectile bombardment of embryogenic suspension cells in *Festuca arundinacea* and *F. rubra*. **A** Suspension cells of tall fescue plated on filter paper disk prior to microprojectile bombardment. **B** β-glucuronidase (GUS)-expressing foci evaluated 15 days post-bombardment of suspension cells from **A**. **C** Hygromycin (Hm)-resistant calli obtained after bombardment of embryogenic suspension cells of tall fescue with pAcH1-coated particles; bombarded cells were subjected to selection with Hm in liquid AAF medium for 4 weeks followed by proliferation on solid MSP medium containing 250mg/l Hm for a further 3 weeks. **D** Shoot differentiation on Hm-resistant callus of tall fescue 4 weeks after transfer onto regeneration medium. **E** Transgenic tall fescue plantlets growing in vitro 14 weeks after microprojectile bombardment of embryogenic cells. **F** Transgenic tall fescue plants growing under greenhouse conditions 5 months after microprojectile bombardment of embryogenic cell suspensions. **G** GUS-expressing foci evaluated 3 days post-bombardment of embryogenic suspension cells of red fescue. **H** Embryogenic suspension cells of red fescue cultured in liquid AAF medium containing 25mg/l G418 after bombardment with pHP23-coated microprojectiles. **I** G418-resistant calli obtained after bombardment of embryogenic suspension cells of red fescue with pHP23-coated particles; bombarded cells were subjected to selection with G418 in liquid AAF medium for 4 weeks followed by proliferation on solid MSP medium containing 50mg/l G418 for 3 weeks. **J** G418-resistant calli of red fescue obtained after bombardment of embryogenic suspension cells with pHP23-coated microprojectiles and selection of bombarded cells for 5 weeks on MSP medium supplemented with 50mg/l G418. **K** Regenerating G418-resistant calli of red fescue 3 weeks after transfer onto regeneration medium. **L** Transgenic soil-grown red fescue plants 5 months after microprojectile bombardment of embryogenic suspension cells

Stable transformation experiments of tall and red fescue cells were performed with *hph* and *npt2* gene expression vectors driven by the CaMV 35S/19S promoter or rice *act1* 5' regulatory sequences. Risk assessments of selectable marker genes have to date focused largely on *npt2*. It has been concluded that ingestion of genetically engineered plants producing neomycin phosphotransferase due to the expression of *npt2* gene poses no safety concerns (Fuchs et al. 1993). After delivery of microprojectiles coated with plasmid DNA bearing the *hph* or *npt2* chimeric genes, the bombarded cells were subjected to selection schemes with hygromycin (50–250 mg/l) and G418 (25–75 mg/l), respectively, and finally led to the recovery of antibiotic-resistant calli (Fig. 8.1C, H–J). Two selection schemes using either ab initio solid plate selection or selection in liquid culture medium and involving a stepwise increase in the antibiotic concentration and subsequent plating on solidified medium were tested. The liquid selection scheme gave the highest recovery frequency, also on the basis of number of bombarded dishes containing antibiotic-resistant calli (Spangenberg et al. 1995a). Using this selection scheme, on average 1.2 and 2.2 antibiotic-resistant calli were recovered per bombarded dish in tall and red fescue, respectively. On the basis of fresh weight of plated cells per experiment, these transformation rates are twice as high as previously reported ones after direct gene transfer to protoplasts of *F. arundinacea* (Wang et al. 1992) and *F. rubra* (Spangenberg et al. 1994; Chap. 7).

Antibiotic-resistant calli regenerated multiple green shoots in 35 and 85% of the cases for tall fescue and red fescue, respectively (Fig. 8.1D, K). Three months after microprojectile bombardment of embryogenic suspension cells, in vitro-rooted plantlets regenerated from the antibiotic-resistant calli (Fig. 8.1E) could be transferred to soil. Transformed tall and red fescue plants

Fig. 8.2A–E. Molecular analysis of transgenic plants of *Festuca arundinacea* and *F. rubra* from microprojectile bombardment of embryogenic suspension cells. **A** Plasmid maps of chimeric *hph* gene construct pAcH1 and *npt2* gene construct pHP23 used for biolistic transformation. **B** Southern blot hybridization with undigested (*odd-numbered lanes*) and *Bam*HI-digested (*even-numbered lanes*) total cellular DNA samples isolated from Hm-resistant tall fescue plants hybridized with an *hph* probe. *Lanes 1–14* Seven independent Hm-resistant regenerants; *15, 16* untransformed tall fescue plant as negative control; *17* 10 pg of *Bam*HI-digested pAcH1 as positive control. **C** Northern blot hybridization with total cellular RNA samples isolated from Hm-resistant tall fescue plants hybridized with an *hph* probe. *Lanes 1–8* Nine independent Hm-resistant transformants of tall fescue; *9* untransformed tall fescue plant as negative control; *10* positive control. **D** Southern blot hybridization with undigested (*odd-numbered lanes*) and *Bam*HI-digested (*even-numbered lanes*) total cellular DNA samples isolated from G418-resistant red fescue plants hybridized with an *npt2* probe. *Lanes 1–8* Four independent G418-resistant regenerants; *9* non-transformed red fescue negative control; *10, 11* 5 pg and 10 pg of *Hin*dIII-digested pHP23 as positive controls, respectively. **E** Northern blot hybridization with total cellular RNA samples isolated from G418-resistant red fescue plants hybridized with an *npt2* probe. *Lanes 1–4* Four independent G418-resistant transformants of red fescue; *5* untransformed red fescue plant as negative control; *6* positive control

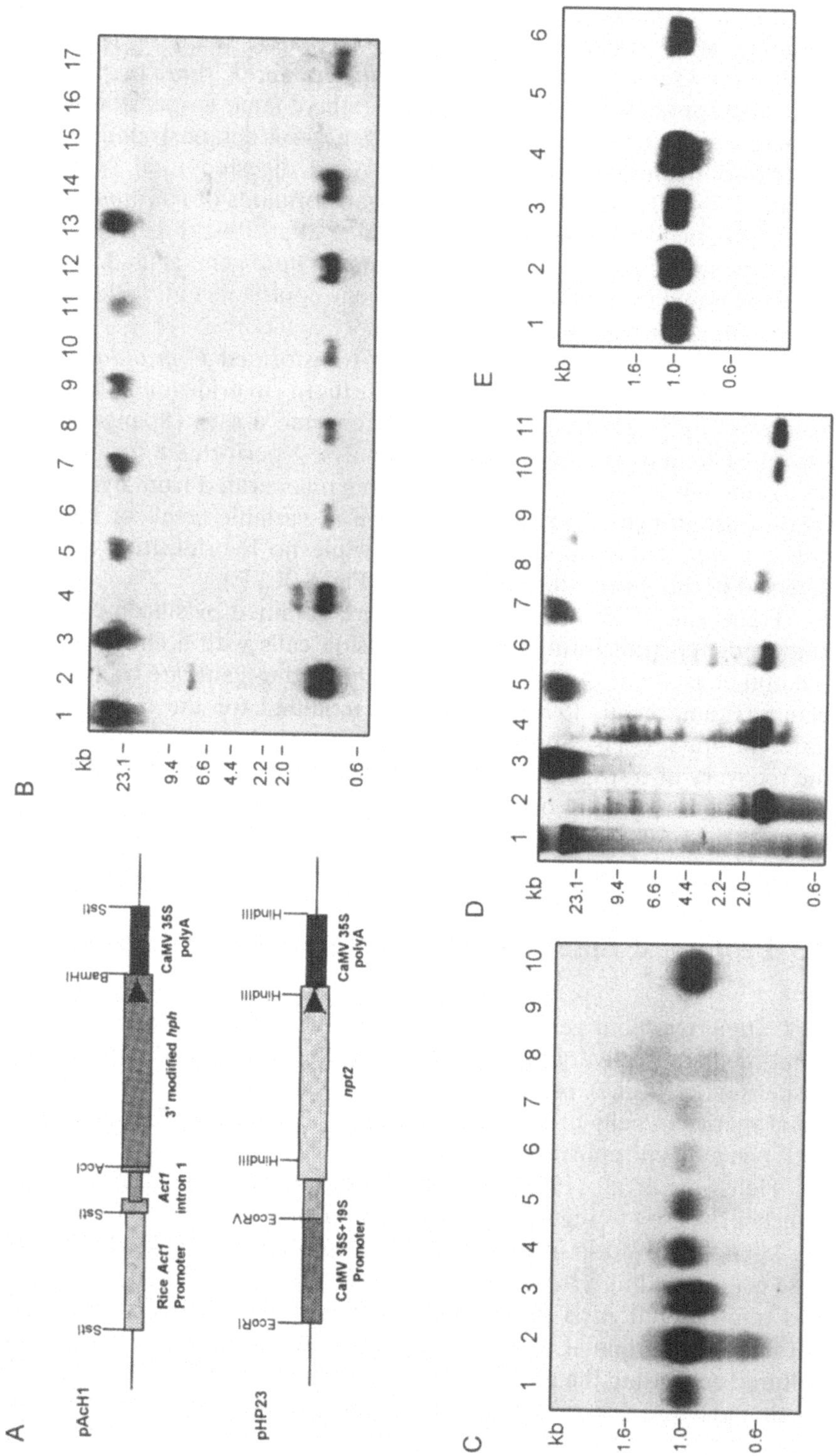
A
pAcH1
SstI
SstI
AccI
BamHI
SstI
Rice Act1 Promoter
Act1 intron 1
3' modified hph
CaMV 35S polyA
pHP23
EcoRI
EcoRV
HindIII
HindIII
HindIII
CaMV 35S+19S Promoter
npt2
CaMV 35S polyA
B
kb
23.1
9.4
6.6
4.4
2.2
2.0
0.6
1 2 3 4 5 6 7 8 9 10 11 12 13 14 15 16 17
C
kb
1.6
1.0
0.6
1 2 3 4 5 6 7 8 9 10
D
kb
23.1
9.4
6.6
4.4
2.2
2.0
0.6
1 2 3 4 5 6 7 8 9 10 11
E
kb
1.6
1.0
0.6
1 2 3 4 5 6

resulting from microprojectile bombardment and grown under greenhouse conditions were obtained within 5 months (Fig. 8.1F, L).

The transgenic nature of the *F. arundinacea* and *F. rubra* plants recovered from microprojectile bombardment of embryogenic suspension cells using chimeric *hph* and *npt2* selectable marker genes was demonstrated by Southern hybridization analyses using undigested and digested total cellular DNA samples (Fig. 8.2). Results obtained for transformants of *F. arundinacea* bearing a chimeric 3' modified *hph* gene (Bilang et al. 1991; Fig. 8.2A, B) and *F. rubra* transgenic plants expressing a chimeric *npt2* gene (Fig. 8.2A, D) indicated the integration of one to five transgene copies, including the integration of rearranged or partial copies (Fig. 8.2).

The expression of the transgenes in transformed *F. arundinacea* and *F. rubra* plants was demonstrated by northern hybridization analysis and hygromycin phosphotransferase (HPH) enzyme assays (Spangenberg et al. 1995a). The northern hybridization analyses performed for independent transgenic tall fescue and red fescue plants regenerated from hygromycin- or G418-resistant calli showed accumulation at variable levels of the expected *hph* and *npt2* transcripts, respectively, while no hybridization signals were detected in the untransformed samples (Fig. 8.2C, E).

Transgenic *F. arundinacea* plants were obtained by silicon-carbide whiskers-mediated transformation of suspension cells with a chimeric *hph* gene (Dalton et al. 1998). A protocol previously used to generate transgenic maize plants (Frame et al. 1994) was slightly modified for the transformation of different grasses. The Southern hybridization analysis provided evidence for the recovery of transgenic tall fescue plants carrying a single copy of the *hph* gene as well as for the ocurrence of transgene rearrangements on insertion (Dalton et al. 1998).

8.3 Protoplast-Independent Transformation in *Lolium*

In *Lolium*, transient gene expression studies after microprojectile bombardment-mediated delivery of foreign DNA into seedlings and non-morphogenic suspension cells of *L. perenne* (Hensgens et al. 1993) and into vegetative and floral meristem cells of *L. multiflorum* and *L. perenne* (Pérez-Vicente et al. 1993) have been reported.

Hensgens et al. (1993) studied expression of chimeric *gusA* gene constructs driven by 5' regulatory sequences of the constitutively expressed rice *gos2* gene involved in initiation of translation and of the light-inducible rice *gos5* gene encoding a 9.5-kDa photosystem I polypeptide. The transcriptional and translational *gusA* gene fusions were tested after microprojectile bombardment of perennial ryegrass seedlings and non-morphogenic suspension-cultured cells using the gun powder particle bombardment device. Expression levels and patterns were compared with those from chimeric *gusA* genes under control of the CaMV 35S promoter and the mannopine synthase (*mas*) promoters 1' and 2' of the T_R-DNA (Hensgens et al. 1993). For the *gos2-gusA* chimeric gene, levels of GUS activity comparable to those obtained with the CaMV 35S promoter-driven *gusA* gene and higher than with the *mas-gusA* chimeric genes were found. Few GUS expressing cells were observed in leaf, root and coleoptile tissues of *gos2-gusA* bombarded ryegrass seedlings as well as in suspension cells. The *gos5-gusA* chimeric gene was found to be active only in green tissues of bombarded ryegrass seedlings.

The direct delivery of microprojectiles carrying plasmid DNA into meristematic cells of vegetative and floral apices of ryegrasses was studied by Pérez-Vicente et al. (1993) as a first step towards establishing conditions of in situ gene transfer to cells possibly contributing to parts of the plant body and the gametes. Excised vegetative apices of *L. perenne* and *L. multiflorum* (Chap. 2) were bombarded with microprojectiles coated with a chimeric *act1-gusA* gene using the microtargeting apparatus developed by Sautter et al. (1991). The microtargeted vegetative apices consisted of the meristematic dome plus two to six pairs of leaf primordia. The conditions applied for propelling the particle suspension (6–8 mm working distance, 110 bar compressed nitrogen) allowed particles to penetrate the first (L1) and second (L2) cell layers in vegetative ryegrass meristems. The number of meristematic L1 and L2 cells expressing the chimeric *gusA* gene per microtargeted vegetative apex varied between one and ten. The delivery of particles to shoot apical ryegrass meristems with the microtargeting device was found not to affect their in vitro culture performance (Pérez-Vicente et al. 1993).

Floral meristems from excised immature inflorescences of Italian ryegrass were also used for microtargeting-mediated delivery of a chimeric *gusA* gene (Pérez-Vicente et al. 1993). Early immature inflorescences with exposed spikelet buds and subtending collar primordia proved to be appropriate developmental stages for delivering particles to flower meristematic cells when laterally microtargeted. Spikelet primordia from these early immature inflorescences were previously shown to differentiate florets with mature anthers containing viable pollen (Chap. 2). Penetration of particles transporting the chimeric *act1-gusA* gene at least to the first and second cell layers of floral organ primordia was demonstrated by corresponding histochemical transient GUS assays (Pérez-Vicente et al. 1993).

Following the studies on transient gene expression by Hensgens et al. (1993), stable transformation and long-term expression of the chimeric *gos2-gusA* gene in callus lines of *L. perenne* obtained by microprojectile bombardment of non-morphogenic suspension cells was reported (van der Maas et al. 1994). A CaMV 35S-driven *hph* gene was used as a selectable marker for these experiments. Bombarded perennial ryegrass suspension cells were selected on solidified medium containing 150 mg/l hygromycin. On average, 5.5 hygromycin-resistant calli were obtained from 250 mg of fresh-weight bombarded suspension cells. The stable integration of one to up to five *gusA* and *hph* gene copies in the genome of the hygromycin-resistant perennial ryegrass

A
B
C
D
E
F
G
H

callus lines was demonstrated by Southern hybridization analyses using digested and undigested DNA samples. GUS activity was detected in 40% of the callus lines after 1 year in culture. Since non-morphogenic suspension cultures were used for biolistic transformation, the recovery of transgenic perennial ryegrass plants was precluded in this study (van der Maas et al. 1994).

Transgenic forage- and turf-type ryegrass plants have been obtained for the first time by microprojectile bombardment of single genotype-derived embryogenic suspension cells of *L. perenne* (Spangenberg et al. 1995b) and *L. multiflorum* (Ye et al. 1997). A chimeric *act1-hph* gene was used as a selectable marker in these studies. Parameters for the bombardment of the ryegrass embryogenic suspension cells with the particle inflow gun were improved using transient expression assays of chimeric *gusA* genes under control of the CaMV 35S and maize ubiquitin (*ubi1*) promoters. The stable transformation experiments with the chimeric *act1-hph* gene expression vector were performed using microprojectile bombardment parameters and treatments of target cells previously optimized on the basis of transient *gusA* gene expression assays. Experimental conditions were essentially the same as those described for the biolistic transformation of tall and red fescues (Spangenberg et al. 1995a). Selection in liquid culture medium involving a stepwise increase in hygromycin concentration (from 50 to 200 mg/l for perennial ryegrass and from 25 to 100 mg/l for Italian ryegrass) and a doubling of the selection pressure in every 2-week subculture step with subsequent plating of bombarded ryegrass suspension cells on selective solidified medium (Fig. 8.3A–C) gave the highest recovery frequency of resistant calli (Spangenberg et al. 1995b; Ye et al. 1997). For perennial ryegrass, on average one hygromycin-resistant callus was recovered in 26% of the bombarded dishes each supporting 200 to 250 mg of fresh-weight embryogenic suspension cells (Fig. 8.3C; Spangenberg et al. 1995b). These transformation frequencies are lower than those reported for microprojectile bombardment of non-morphogenic *L. perenne* cell suspensions by van der Maas et al. (1994). For Italian ryegrass,

◀

Fig. 8.3A–H. Transgenic *Lolium perenne* plants from particle inflow gun-mediated microprojectile bombardment of embryogenic suspension cells. **A** Suspension cells of perennial ryegrass plated on filter paper disk prior to microprojectile bombardment. **B** Embryogenic suspension cells of perennial ryegrass cultured in liquid AAF medium containing 50 mg/l hygromycin (Hm) 2 weeks after bombardment with pAcH1-coated microprojectiles. **C** Hm-resistant calli obtained after bombardment of embryogenic suspension cells of perennial ryegrass with pAcH1-coated particles; bombarded cells were subjected to selection with Hm in liquid AAF medium for 4 weeks followed by proliferation on solid MSP medium containing 200 mg/l Hm for 3 weeks. **D** Shoot differentiation on Hm-resistant callus of perennial ryegrass 4 weeks after transfer onto regeneration medium. **E** Transgenic perennial ryegrass plantlets growing in vitro 14 weeks after microprojectile bombardment of embryogenic cells. **F** Transgenic soil-grown perennial ryegrass plants 5 months after bombardment of embryogenic suspensions. **G** Inflorescence of transgenic perennial ryegrass plant with protruding anthers. **H** Lugol-stained pollen collected from transgenic plants of perennial ryegrass. (Spangenberg et al. 1995b)

higher transformation frequencies with the average recovery of one resistant callus in 59% of the bombarded dishes were obtained (Ye et al. 1997).

Upon transfer to regeneration medium, 23% of the perennial ryegrass resistant calli produced multiple green shoots and regenerated in vitro-rooted plantlets (Fig. 8.3D, E; Spangenberg et al. 1995b). For Italian ryegrass, in vitro-growing plantlets were regenerated from 33% of the hygromycin-resistant calli obtained (Ye et al. 1997). Regeneration frequencies from transformed ryegrass calli were in the range of those observed for non-bombarded embryogenic cell suspensions of the same genotypes, thus indicating that selection with hygromycin for up to 8 weeks did not impair morphogenesis. Transformed perennial and Italian ryegrass plants from microprojectile bombardment were established in soil and grown under greenhouse conditions (Fig. 8.3F). Transgenic perennial ryegrass plants were vernalized, flowered and produced viable pollen (Fig. 8.3G, H).

The transgenic nature of the *L. perenne* and *L. multiflorum* plants recovered from microprojectile bombardment of embryogenic suspension cells was demonstrated by Southern hybridization analyses using undigested and digested total cellular DNA samples (Fig. 8.4A; Spangenberg et al. 1995b; Ye et al. 1997). Evidence for the integration of single and multiple copies of the *hph* gene in the plant genome was observed for both, transformed perennial ryegrass (Fig. 8.4A) and Italian ryegrass (Ye et al. 1997). The integration of functional transgene copies was demonstrated by northern hybridization analyses and HPH enzyme assays of transgenic perennial ryegrass (Fig. 8.4B, C; Spangenberg et al. 1995b) and Italian ryegrass (Ye et al. 1997) plants.

The meiotic transmission of transgenes was studied in selected transgenic Italian ryegrass plants after reciprocal crossings with wild type untransformed control plants (Ye 1997). Expression of transgenes in pollen grains from flowering Italian ryegrass plants co-transformed with chimeric *act1-hph* and

Fig. 8.4A–C. Molecular analysis of transgenic plants of *Lolium perenne* from microprojectile bombardment of embryogenic suspension cells. **A** Southern blot hybridization with undigested (*odd-numbered lanes*) and *Bam*HI-digested (*even-numbered lanes*) total cellular DNA samples isolated from perennial ryegrass plants derived from microprojectile bombardment and hybridized with the *hph* probe. *Lanes 1–12* Six independent pAcH1-transformed perennial ryegrass plants; *13, 14* non-transformed negative control plants; *15* 20 pg of *Bam*HI-digested pAcH1 as positive control. **B** Northern blot hybridization using total cellular RNA samples isolated from leaves of perennial ryegrass plants derived from microprojectile bombardment and hybridized with the *hph* probe. *Lanes 1–6* Six independent pAcH1-transformed plants of perennial ryegrass (same plants as in **A**); *7* untransformed perennial ryegrass plant as negative control; *8* 50 pg of *Bam*HI-digested pAcH1 as positive control. **C** Expression of hygromycin phosphotransferase (HPH) in transgenic plants of perennial ryegrass derived from microprojectile bombardment of embryogenic suspension cells. *Lanes 1–8* Four independent pAcH1-transformed plants of perennial ryegrass; *9, 10* untransformed perennial ryegrass plant as negative control; *11, 12* transformed Hm-resistant tobacco plant as positive control. Assays were performed in the presence (*odd-numbered lanes*) and absence (*even-numbered lanes*) of hygromycin. (Spangenberg et al. 1995b)

gusA genes was shown by histochemical GUS staining (Fig 8.5A). Seeds harvested from the reciprocal crosses were germinated in presence of 200 mg/l hygromycin to assess the meiotic transmission of functional *hph* transgene copies. Half of the T_1 seedlings showed vigorous growth in presence of the selective agent (hygromycin-resistant seedlings), while germination of the remaining seeds was strongly inhibited and the primary roots of the correspond-

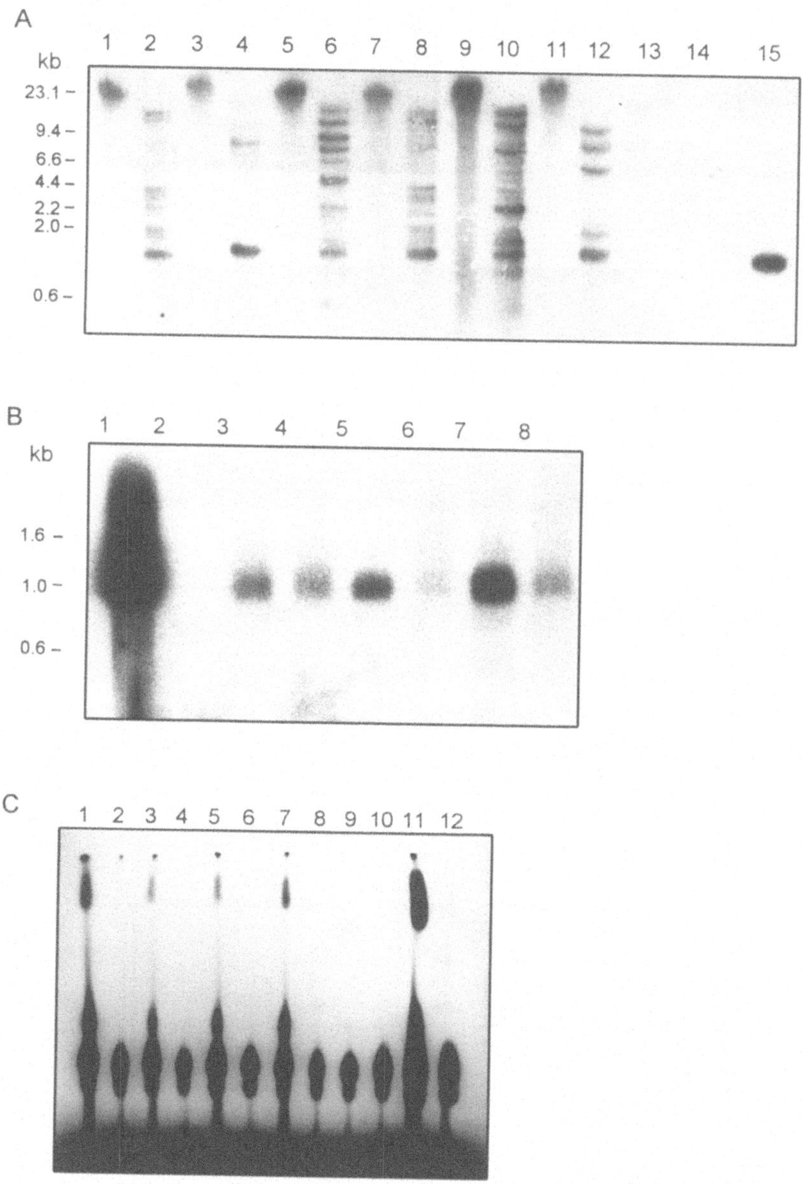

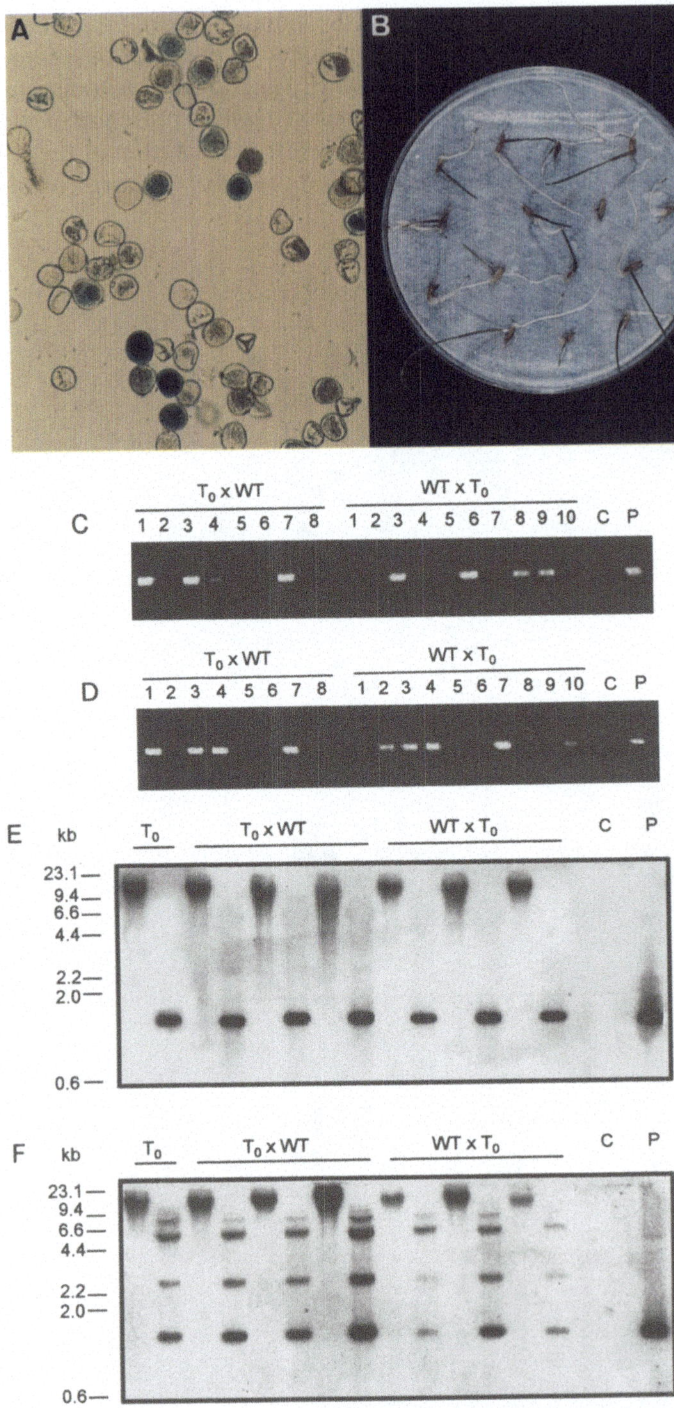
A
B
C
T0 x WT
1 2 3 4 5 6 7 8
WT x T0
1 2 3 4 5 6 7 8 9 10
C P
D
T0 x WT
1 2 3 4 5 6 7 8
WT x T0
1 2 3 4 5 6 7 8 9 10
C P
E
kb
T0
T0 x WT
WT x T0
C P
23.1
9.4
6.6
4.4
2.2
2.0
0.6
F
kb
T0
T0 x WT
WT x T0
C P
23.1
9.4
6.6
4.4
2.2
2.0
0.6

ing stunted seedlings (hygromycin-sensitive seedlings) became necrotic (Fig. 8.5B). Results from a polymerase chain reaction (PCR) screening of T_1 seedlings for detecting the presence of the *hph* transgene were consistent with those from the hygromycin-resistance screening (Fig. 8.5C, D). These results were confirmed by Southern hybridization analysis, where all *hph*-PCR positive T_1 plants showed the same hybridization patterns as the corresponding T_0 parents (Fig. 8.5E, F). Expected 1:1 segregation ratios of *hph* transgene transmission were observed in screened T_1 seedlings from independent T_0 transgenic Italian ryegrass plants. These results unequivocally demonstrated the stable meiotic transmission of the *hph* transgene following Mendelian rules in transgenic ryegrass plants for the first time.

Few independent transgenic *L. multiflorum* plants and a single transgenic *L. perenne* plant were recovered after whiskers-mediated transformation of suspension cells with a chimeric *hph* gene following hygromycin selection (Dalton et al. 1998). The plants recovered were subjected to PCR and Southern hybridization analysis to confirm their transgenic nature. Evidence on the unstable paternal transmission of a functional *hph* gene was provided for two *L. multiflorum* transgenic plants (Dalton et al. 1998).

8.4 Protoplast-Independent Transformation in Other Grasses

Protoplast-independent DNA delivery methods have been used for transient gene expression studies and for the recovery of transgenic plants in *Pennisetum glaucum* (Taylor and Vasil 1991), *Agrostis alba* (Asano et al. 1991), *A. palustris* (Zhong et al. 1993; Hartman et al. 1994; Xiao and Ha 1997), *A. stolonifera* (Dalton et al. 1998) and *Dactylis glomerata* (Denchev et al. 1997).

Physical parameters affecting expression of a chimeric *gusA* gene driven by an enhanced CaMV 35S promoter following microprojectile bombardment of isolated immature zygotic embryos of *P. glaucum* were investigated as a first step towards transformation of pearl millet (Taylor and Vasil 1991). The DuPont PDS-1000 biolistic device was used in this study. Strongest GUS-expression was found in scutellar cells located in the centre of GUS-expressing

Fig. 8.5A–F. Meiotic transmission of transgenes in *Lolium multiflorum* transgenic plants. **A** A *35S-gusA* chimeric gene expressed in pollen grains of T_0 transgenic *L. multiflorum* plants. **B** T_1 seeds from transgenic *L. multiflorum* plant germinated on 200 mg/l hygromycin. **C,D** Polymerase chain reaction (PCR) screening with *hph* primers of DNA samples from T_1 seedlings obtained after reciprocal crossings of two independent transgenic T_0 *L. multiflorum* plants with untransformed Italian ryegrass. **E,F** Southern hybridization analysis of DNA samples from PCR positive T_1 plants from **C** and **D**; DNA was either undigested or digested with *Bam*HI and probed with 1.3 kb *hph* DNA fragment. (Ye 1997)

foci from bombarded immature zygotic embryos *of P. glaucum* subjected to histochemical GUS assays. The surrounding cells showed decreased intensity with increasing distance from the central cells. GUS-expressing cells were found in up to 12 cell layers beneath the epidermis. On average seven GUS-expressing foci per bombarded zygotic embryo were observed. Maximum transient *gusA* gene expression was found to be associated with most tissue damage caused by the formation of pits or by shock damage affecting numerous cell layers (Taylor and Vasil 1991). The bombardment parameters tested (speed and bombardment distance) had limited influence on embryogenic callus formation and plant regeneration from bombarded immature zygotic embryos of pearl millet. No transgenic plant was recovered.

Transient GUS expression was assessed after silicon-carbide whiskers-mediated transformation of intact suspension cells of *A. alba* by Asano et al. (1991). A chimeric CaMV 35S-driven *gusA* gene borne on standard or replicative wheat dwarf virus (WDV)-derived vectors was used. The highest number of GUS-expressing foci was observed when suspension cells in the exponential growth phase were used for transfection. GUS expression levels were higher when the replicative *gusA* vector was transfected. Up to 23 GUS-expressing foci per 250μl of packed cell volume sample of transfected suspension cells were detected (Asano et al. 1991).

Transgenic *A. palustris* plants were first obtained by microprojectile bombardment of the embryogenic callus using the PDS-1000 biolistic particle acceleration device and a chimeric *act1-gusA* gene construct and by subsequent GUS screening (Zhong et al. 1993). GUS activity was detected on average in 15% of the embryogenic callus lines recovered. However, of the 500 plants regenerated, only 4 had integrated the foreign gene, as demonstrated by Southern hybridization analysis (Zhong et al. 1993). Three transgenic plants showed integration of a single transgene copy and one plant had three *gusA* gene copies integrated per haploid genome. One transgenic creeping bentgrass plant was analyzed in detail. The expression of the *gusA* gene was demonstrated by northern and western hybridization analyses. Histochemical assays showed GUS activity in all plant organs tested, with the strongest activity found in stem nodes, the first young leaf and the root tip.

The reproducible generation of transgenic creeping bentgrass plants from different commercial cultivars by biolistic transformation has been reported by Hartman et al. (1994). Chimeric *bar* genes under control of a CaMV 35S promoter were used as selectable markers. Embryogenic suspension-cultured cells were used as targets for microprojectile bombardment with the PDS 1000 biolistic device. Bombarded cells were subjected to plate or liquid selection on media containing 2 or 4 mg/l bialaphos. Approximately 900 plantlets were recovered, but only 55 survived the spraying of supralethal doses of the commercial herbicide Herbiace, the remainder being escapes from the selection schemes. The frequency of plants recovered after selection to survive the high-rate application of the herbicide ranged from 0 to 14%. The presence of the *bar* gene was demonstrated by PCR and Southern hybridiyation analysis in the bialaphos resistant plants. Expression of the transgene was further shown by northern hybridization analysis. Field evaluation of *bar*-transgenic creeping bentgrass plants revealed a high level of resistance to applications of the threefold field rate of the herbicide (Lee 1996). Efficient selection with 200 mg/l hygromycin allowed for the recovery of transgenic *A. palustris* plants after biolistic transformation of callus and suspension cells using a chimeric *hph* gene and the PDS 1000 biolistic device (Xiao and Ha 1997).

A few transgenic *A. stolonifera* plants, likely derived from a single transformation event, were obtained after silicon-carbide whiskers-mediated transformation of suspension cells with a chimeric *hph* gene (Dalton et al. 1998).

On the basis of a unique direct somatic embryogenesis system from leaf explants available for *Dactylis glomerata* (Conger et al. 1986; Trigiano et al. 1989), transgenic somatic embryos and plants expressing chimeric *gusA* and *bar* genes have been recovered after particle inflow gun-mediated transformation of leaf base cells (Denchev et al. 1997). Unless such regeneration systems are established for other forage and turf grasses, embryogenic callus and suspension cells will remain the most frequent target for biolistic transformation.

8.5 Summary and Conclusions

Within the *Festuca-Lolium* complex, proven transgenic plants have been generated by protoplast-independent methods so far in *F. arundinacea*, *F. rubra*, *L. perenne* and *L. multiflorum*. The transgenic plants were mainly obtained by biolistic transformation of single genotype-derived embryogenic suspension cells using the cost-effective particle inflow gun. Genes conferring resistance to the antibiotics hygromycin and geneticin were chosen as selectable markers, and tight selection schemes were developed. The transgenic character of the regenerated plants was considered to be proven following the same criteria as outlined in Section 7.5 (Chap. 7). Although many of the transgenic fescues and ryegrasses produced by microprojectile bombardment technology still show complex transgene integration patterns, transgenic plants with single copy integration were also obtained. The meiotic stable maternal and paternal transmission of transgenes was unequivocally demonstrated so far for transgenic *L. multiflorum* plants obtained from biolistic transformation only.

Whiskers-mediated transformation – another protoplast-independent gene transfer method – has led so far to the recovery of few transgenic plants in *F. arundinacea* and *L. multiflorum*, and a single transgenic plant in *L. perenne* and in *A. stolonifera*. These outcomes and constraints in treating large number of cells with current protocols suggest that the whiskers-mediated transformation may not represent the expected inexpensive alternative to biolistic transformation. In contrast, recent success in the generation of transgenic rice, maize, wheat and barley by *Agrobacterium*-mediated transformation (Hiei et al. 1994; Ishida et al. 1996; Rashid et al. 1996; Cheng et al. 1997; Tingay et al. 1997) encourages the evaluation of *A. tumefaciens* as a vector for and highlight its potential in producing transgenic forage and turf grass plants.

References

Asano Y, Otsuki Y, Ugaki M (1991) Electroporation-mediated and silicon carbide fiber-mediated DNA delivery in *Agrostis alba* L. (redtop). Plant Sci 79: 247–252

Batty NP, Evans JM (1992) Biological ballistics – no longer a shot in the dark. Transgenic Res 1: 107–113

Bilang R, Iida S, Peterhans A, Potrykus I, Paszkowski J (1991) The 3′ terminal region of the hygromycin-B-resistance gene is important for its activity in *Escherichia coli* and *Nicotiana tabacum*. Gene 100: 247–250

Casas AM, Kononowicz AK, Zehr UB, Tomes DT, Axtell JD, Butler LG, Bressan RA, Hasegawa PM (1993) Transgenic sorghum plants via microprojectile bombardment. Proc Natl Acad Sci USA 90: 11212–11216

Cheng M, Fry JE, Pang S, Zhou H, Hironaka CM, Duncan DR, Conner TW, Wan Y (1997) Genetic transformation of wheat mediated by *Agrobacterium tumefaciens*. Plant Physiol 115: 971–980

Christou P (1992) Genetic transformation of crop plants using microprojectile bombardment. Plant J 2: 275–281

Christou P (1993) Particle gun-mediated transformation. Curr Opin Biotechnol 4: 135–141

Christou (1996) Electric discharge particle acceleration (Accell®) technology for the creation of transgenic plants with altered characteristics. Field Crops Res 45: 143–151

Christou P, Ford T, Kofron M (1991) Production of transgenic rice (*Oryza sativa* L.) plants from agronomically important Indica and Japonica varieties via electric discharge particle acceleration of exogenous DNA into immature zygotic embryos. Bio/Technology 9: 957–962

Conger BV, Hanning GE, Gray DJ, McDaniel JK (1983) Direct embryogenesis from mesophyll cells of orchardgrass. Science 221: 850–851

Dalton SJ, Bettany AJE, Timms E, Morris P (1998) Transgenic plants of *Lolium multiflorum*, *Lolium perenne*, *Festuca arundinacea* and *Agrostis stolonifera* by silicon carbide fibre-mediated transformation of cell suspension cultures. Plant Sci (in press)

Denchev PD, Songstad DD, McDaniel JK, Conger BV (1997) Transgenic orchardgrass (*Dactylis glomerata*) plants by direct embryogenesis from microprojectile bombarded leaf cells. Plant Cell Rep 16: 813–819

Dunder E, Dawson J, Suttie J, Pace G (1995) Maize transformation by microprojectile bombardment of immature embryos. In: Potrykus I, Spangenberg G (eds) Gene transfer to plants. Springer, Berlin Heidelberg New York, pp 127–138

Dunahay TG (1993) Transformation of *Chlamydomonas reinhardtii* with silicon carbide whiskers. BioTechniques 15: 452–460

Finer JJ, Vain P, Jones MW, McMullen MD (1992) Development of the particle inflow gun for DNA delivery to plant cells. Plant Cell Rep 11: 323–328

Frame BR, Drayton PR, Bagnall SV, Lewnau CJ, Bullock WP, Wilson HM, Dunwell JM, Thompson JA, Wang K (1994) Production of fertile transgenic maize plants by silicon carbide whisker-mediated transformation. Plant J 6: 941–948

Fuchs RL, Ream JE, Hammond BG, Naylor MW, Leimgruber RM, Berberich SA (1993) Case study in safety assessment (II): safety assessment of the neomycin phosphotransferase II (*nptII*) protein. Bio/Technology 11: 1543–1547

Hartman CL, Lee L, Day PR, Tumer NE (1994) Herbicide resistant turfgrass (*Agrostis palustris* Huds.) by biolistic transformation. Bio/Technology 12: 919–923

Hensgens LAM, de Bakker EPHM, van Os-Ruygrok EP, Rueb S, van de Mark F, van der Maas HM, van der Veen S, Kooman-Gersmann M, Hart L, Schilperoort RA (1993) Transient and stable expression of *gusA* fusions with rice genes in rice, barley and perennial ryegrass. Plant Mol Biol 22: 1101–1127

Hiei J, Ohta S, Komari T, Kumashiro T (1994) Efficient transformation of rice (*Oryza sativa* L.) mediated by *Agrobacterium* and sequence analysis of the boundaries of the T-DNA. Plant J 6: 271–282

Ishida Y, Saito H, Ohta S, Hiei Y, Komari T, Kumashiro T (1996) High efficiency transformation of maize (*Zea mays* L.) mediated by *Agrobacterium tumefaciens*. Bio/Technology 14: 745–750

Kaeppler HF, Gu W, Somers DA, Rines HW, Cockburn AF (1990) Silicon carbide fiber-mediated DNA delivery into plant cells. Plant Cell Rep 9: 415–418

Kaeppler HF, Somers DA, Rines HW, Cockburn AF (1992) Silicon carbide fiber-mediated stable transformation of plant cells. Theor Appl Genet 84: 560–566

Klein TM (1995) The biolistic transformation system. In: Potrykus I, Spangenberg G (eds) Gene transfer to plants. Springer, Berlin Heidelberg New York, pp 115–117

Klein TM, Fitzpatrick-McElligott S (1993) Particle bombardment – a universal approach for gene transfer to cells and tissues. Curr Opin Biotechnol 4: 583–590

Klein TM, Wolf ED, Wu R, Sanford JC (1987) High velocity microprojectiles for delivering nucleic acids into living cells. Nature 327: 70–73

Lee L (1996) Turfgrass biotechnology. Plant Sci 115: 1–8

McCabe DE, Christou P (1993) Direct DNA transfer using electric discharge particle acceleration (Accell®) technology. Plant Cell Tissue Organ Culture 33: 227–236

Pérez-Vicente R, Wen XD, Wang ZY, Leduc N, Sautter C, Wehrli E, Potrykus I, Spangenberg G (1993) Culture of vegetative and floral meristems in ryegrasses: potential targets for microballistic transformation. J Plant Physiol 142: 610–617

Potrykus I (1990) Gene transfer to cereals: an assessment. Bio/Technology 8: 535–542

Rashid H, Yokoi S, Toriyama K, Hinata K (1996) Transgenic plant production mediated by *Agrobacterium* in *Indica* rice. Plant Cell Rep 15: 727–730

Sanford JC (1988) The biolistic process. TIBTECH 6: 299–302

Sanford JC, Klein TM, Wolf ED, Allen NJ (1987) Delivery of substances into cells and tissues using a particle bombardment process. J Particulate Sci Technol 6: 559–563

Sanford JC, Smith FD, Russell JA (1993) Optimizing the biolistic process for different biological applications. Methods Enzymol 217: 483–509

Sautter C, Waldner H, Neuhaus-Url G, Galli A, Neuhaus G, Potrykus I (1991) Micro-targeting: high efficiency gene transfer using a novel approach for the acceleration of microprojectiles. Bio/Technology 9: 1080–1085

Sautter C (1993) Development of a microtargeting device for particle bombardment of plant meristems. Plant Cell Tissue Organ Cult 33: 251–257

Spangenberg G, Wang ZY, Nagel J, Potrykus I (1994) Protoplast culture and generation of transgenic plants in red fescue (*Festuca rubra* L.). Plant Sci 97: 83–94

Spangenberg G, Wang ZY, Wu XL, Nagel J, Iglesias VA, Potrykus I (1995a) Transgenic tall fescue (*Festuca arundinacea*) and red fescue (*F. rubra*) plants from microprojectile bombardment of embryogenic suspension cells. J Plant Physiol 145: 693–701

Spangenberg G, Wang ZY, Wu XL, Nagel J, Potrykus I (1995b) Transgenic perennial ryegrass (*Lolium perenne*) plants from microprojectile bombardment of embryogenic suspension cells. Plant Sci 108: 209–217

Spangenberg G, Wang ZY, Potrykus I (1998) Biotechnology in forage and turf grass improvement. In: Frankel R, Grossman M, Linskens HF, Maliga P, Riley R (eds) Monographs on theoretical and applied genetics, this volume. Springer, Berlin Heidelberg New York

Svab Z, Maliga P (1993) High-frequency plastid transformation in tobacco by selection for a chimeric *aad*A gene. Proc Natl Acad Sci USA 90: 913–917

Svab Z, Hajdukiewicz P, Maliga P (1990) Stable transformation of plastids in higher plants. Proc Natl Acad Sci USA 87: 8526–8530

Takamizo T, Suginobu KI, Ohsugi R (1990) Plant regeneration from suspension culture derived protoplasts of tall fescue (*Festuca arundinacea* Schreb.) of a single genotype. Plant Sci 72: 125–131

Taylor MG, Vasil IK (1991) Histology of, and physical factors affecting transient GUS expression in pearl millet (*Pennisetum glaucum* (L.) R. Br.) embryos following microprojectile bombardment. Plant Cell Rep 10: 120–125

Tingay S, McElroy D, Kalla R, Fieg S, Wang M, Thornton S, Brettell R (1997) *Agrobacterium tumefaciens*-mediated barley transformation. Plant J 11: 1369–1376

Trigiano RN, Gray DJ, Conger BV, McDaniel JK (1989) Origin of direct somatic embryos from cultured leaf segments of *Dactylis glomerata*. Bot Gaz 150: 72–77

Vain P, Keen N, Murillo J, Rathus C, Nemes C, Finer JJ (1993) Development of the particle inflow gun. Plant Cell Tissue Organ Cult 33: 237–246

Van der Maas HM, de Jong ER, Rueb S, Hensgens LAM, Krens FA (1994) Stable transformation and long-term expression of the *gusA* reporter gene in callus lines of perennial ryegrass (*Lolium perenne* L.). Plant Mol Biol 24: 401–405

Vasil IK (1988) Progress in the regeneration and genetic manipulation of cereal crops. Bio/Technology 6: 397–402

Vasil V, Srivastava V, Castillo AM, Fromm ME, Vasil IK (1993) Rapid production of transgenic wheat plants by direct bombardment of cultured immature embryos. Bio/Technology 11: 1553–1558

Wan Y, Lemaux PG (1994) Generation of large numbers of independently transformed fertile barley plants. Plant Physiol 104: 37–48

Wang ZY, Takamizo T, Iglesias VA, Osusky M, Nagel J, Potrykus I, Spangenberg G (1992) Transgenic plants of tall fescue (*Festuca arundinacea* Schreb.) obtained by direct gene transfer to protoplasts. Bio/Technology 10: 691–696.

Wang K, Frame BR, Drayton PR, Thompson JA (1995) Silicon carbide whisker-mediated transformation: regeneration of transgenic maize plants. In: Potrykus I, Spangenberg G (eds) Gene transfer to plants. Springer, Berlin Heidelberg New York, pp 186–192

Xiao L, Ha SB (1997) Efficient selection and regeneration of creeping bentgrass transformants following particle bombardment. Plant Cell Rep 16: 874–878

Ye XD (1997) Gene transfer to ryegrasses (*Lolium* spp.): modification of fructan metabolism in transgenic plants. PhD Diss, Swiss Federal Institute of Technology, Zürich

Ye XD, Wang ZY, Wu XL, Potrykus I, Spangenberg G (1997) Transgenic Italian ryegrass (*Lolium multiflorum*) plants from microprojectile bombardment of embryogenic suspension cells. Plant Cell Rep 16: 379–384

Zhong H, Bolyard MG, Srinivasan C, Stickelen M (1993) Transgenic plants of turfgrass (*Agrostis palustris* Huds.) from microprojectile bombardment of embryogenic callus. Plant Cell Rep 13: 1–6

Chapter 9

Molecular Markers

9.1 Introduction

The development of molecular markers is based on naturally occurring polymorphism which has been shown to be widespread in natural populations and to differ markedly in different species (Evola et al. 1986; Burr 1994). Polymorphism can be revealed at different levels depending on the detection method used.

About 50% of all enzymes investigated so far exist in multiple molecular forms called isozymes (Rothe 1994). These enzymically active proteins have the same substrate specificity but usually differ in electrophoretic mobility (McMillin 1983; Rothe 1994). Isozymes were the first markers used for linkage studies and for the construction of genetic maps of plant chromosomes (Moore and Collins 1983; Helentjaris 1992). Many of the principles applicable to molecular markers were conceptualized during these early studies on isozymes (Helentjaris 1992).

Isozymes have been used very successfully in certain aspects of genetics and plant breeding as near-neutral genetic markers, such as in variety identification, seed purity determination and population genetic studies (Brown 1978; Moore and Collins 1983; Nielsen 1985; Tanksley et al. 1989; Greneche et al. 1991). Unfortunately, the number of genetic markers provided by isozyme assays is insufficient for many applications in plant breeding (Tanksley et al. 1989).

The limitation of isozymes has been superseded by the use of DNA-based markers, such as restriction fragment length polymorphisms (RFLPs), polymerase chain reaction (PCR)-based markers, as well as species-specific repetitive DNA sequences. The polymorphisms revealed at the DNA level are much more abundant than those identified at the protein and morphological levels (Tanksley et al. 1989; Helentjaris 1992).

Repetitive DNA sequences contribute considerably to the genome size of higher plants. Species-specific repetitive DNA sequences have been proven useful in studying genome evolution and species divergence at the molecular level (Pérez-Vicente et al. 1992). In addition, they represent valuable tools for analyzing hybrid genomes (Chap. 6).

Restriction endonuclease digestion of total genomic DNA followed by hybridization with a labelled probe reveals differently sized hybridization fragments. This form of polymorphism, termed restriction fragment length polymorphism (RFLP), has been shown to have the advantage of being universal, abundant, stable and convenient (Xu and Sleper 1991). RFLP markers have been studied extensively for genome mapping, germplasm evaluation and marker-aided breeding (Beckmann and Soller 1986; Tanksley et al. 1989; Waugh and Powell 1992).

Random amplified polymorphic DNA (RAPD) is based on PCR amplification of genomic DNA using single random sequence primers (usually 10-mers). It is based on the observation that single short oligonucleotide primers can frequently recognize similar sequences that oppose each other at distances close enough for the intervening sequence to be amplified by PCR (Welsh and McClelland 1990; Williams et al. 1990; Burr 1994). Genetic variations that result from changes in either the sequence of the primer binding site (e.g. point mutations) or from changes which alter the size or prevent the successful amplification of target DNA (e.g. insertions, deletions, inversions) can be revealed (Baird et al. 1992; Waugh and Powell 1992). This class of DNA-based markers has been applied to fingerprinting, gene mapping and population genetic structure studies (Rafalski and Tingey 1993; Rafalski et al. 1994).

In contrast to somatic hybridization (Chap. 6) and genetic transformation (Chaps. 7, 8), molecular markers are not aimed at restructuring genomes, but are service-oriented and basically intended to increase the efficiency of plant improvement by providing additional, more sophisticated and discriminatory tools to plant breeders (Beckmann and Soller 1986).

Studies on the development and application of isozyme and DNA-based markers in forage and turf grasses, especially in fescues and ryegrasses, are reviewed below.

9.2 Molecular Markers in *Festuca* and *Lolium*

9.2.1 Isozyme Markers

One of the most important applications of isozyme markers in grass breeding has been in varietal discrimination, particularly in the genus *Lolium*. Candidate ryegrass cultivars are generally evaluated based on their distinctness, uniformity and stability. Due to similarity in their morphology and the inherent variability within cultivars in cross-pollinated species coupled with the availability of a limited number of suitable phenotypic characters, establishing distinctness in grass cultivars has become increasingly difficult (Gilliland et al. 1982; Nielsen et al. 1985). This has prompted the search for additional identification tools.

The first report on the use of well-defined allelic isozymes to identify ryegrass cultivars was that of Hayward and McAdam (1977). Two enzyme systems, phosphoglucoisomerase (PGI) and glutamate oxaloacetate-transaminase (GOT), were used for the evaluation of eight perennial ryegrass cultivars. Differences between cultivars were revealed by comparing the allele frequencies at three loci – *Got 2*, *Got 3*, and *Pgi 2*. All but two of the 36 possible pair-wise comparisons showed significant differences for one or more loci (Hayward and McAdam 1977). This study clearly revealed the distinctness amongst a set of very closely related cultivars by the examination of just two enzyme systems with three polymorphic loci. This method for distinguishing between outbreeding cultivars is based on the statistical analysis of isozyme frequencies. Therefore it does not require the total exclusion of an isozyme from one cultivar and its presence in another but rather exploits the natural variability in outbreeding cultivars as a basis for identification.

Gilliland et al. (1982) confirmed the usefulness of this technique by examining genotype frequencies for one locus (*Pgi 2*) in 23 perennial ryegrass cultivars. Out of a possible total of 253 pair-wise comparisons, 193 comparisons (76%) showed significant differences. This result was achieved with much less effort and showed discriminating power similar to that normally obtained by field evaluation of phenotypic characters concerning distinctness, uniformity and stability for Plant Breeder Rights (Gilliland et al. 1982).

It is obvious that the use of more loci showing reasonable variation will improve the method considerably. In a survey of 15 perennial ryegrass and 6 Italian ryegrass cultivars, Nielsen et al. (1985) examined allele frequencies for four loci – *Pgi 2*, *Got 3*, *Pgm 1* (phosphoglucomutase) and *Pgd 1* (6-phosphogluconate dehydrogenase). The cultivars were compared two by two for homogeneity of the allele frequencies of each locus. The two loci *Pgi 2* and *Got 3* distinguished perennial ryegrass cultivars fairly well, but the discrimination in Italian ryegrass was unsatisfactory at any single locus. By summing up the test values for each locus, the cultivars could be distinguished pair-wise in 96% of pairs for perennial ryegrass and in 80% of pairs for Italian ryegrass. Furthermore, indications of instability of some cultivars were found from homogeneity tests between seed lots (Nielsen et al. 1985). Greneche et al. (1991) tested six enzymes for distinguishing ryegrass cultivars: acid phosphatase (ACP), isocitrate dehydrogenase (IDH), malate dehydrogenase (MDH), PGI, shikimate dehydrogenase (ShDH) and superoxide dismutase (SOD). PGI, ACP and IDH were found to be efficient enzymes for varietal discrimination. With a combined evaluation of the three enzymes PGI, ACP and IDH, all 16 perennial ryegrass and 10 Italian ryegrass cultivars tested could be distinguished from one another (Greneche et al. 1991).

Isozyme patterns are more complicated in tetraploid than in diploid plants, since a diploid plant has a maximum of two alleles per locus, while a tetraploid plant may contain four different alleles per locus. In a study using tetraploid *L. perenne* and *L. multiflorum* and their hybrids, Nielsen (1980) found a good agreement between the observed and the expected patterns, with all the possible 35 genotypes identified by the *Pgi 2* isozyme pattern. This result demonstrated that the *Pgi 2* enzyme system is an excellent tool for the separation of hybrids and self-pollinations in a breeding program for tetraploid ryegrass. The *Pgi 2* isozyme system was further tested in a survey of 11 cultivars of tetraploid perennial ryegrass and 8 cultivars of tetraploid Italian ryegrass, showing that for both species the cultivars could be well discriminated by the frequencies of four alleles (Østergaard and Nielsen 1981).

Based on an analysis of PGI, ACP and IDH loci which allowed the identification of most cultivars, a catalogue of allelic frequencies for 139 perennial ryegrass and Italian ryegrass cultivars, which were both diploid and tetraploid, was produced (Lallemand et al. 1991). The two species differed by their mean allelic frequencies for all three enzymes tested. Diploid and tetraploid cultivars had slightly different distributions especially for PGI among Italian ryegrasses (Lallemand et al. 1991). The variation of allele frequencies in the catalogue provides a tool to determine whether a candidate cultivar is genetically different from any of the existing cultivars, enables the identity and stability of seed lots to be checked, and allows the composition of commercial mixtures to be controlled.

Considerably less effort has been made in the use of isozymes for cultivar identification in tall fescue, since its allopolyploid nature causes difficulties in genetic analysis. Hicks et al. (1982) used peroxidase isozyme to analyze three tall fescue cultivars and four allopolyploid accessions. Little isozyme variation was observed in the tetraploid and hexaploid, but considerably more in the octoploid and decaploid plants. No consistent differences were found in band number, position or frequencies among the cultivars tested in spite of their diverse parentage. The lack of unique bands seems to rule out peroxidase as an aid for cultivar identification in tall fescue (Hicks et al. 1982).

Isozymes have the potential for revealing somaclonal variation in *Festuca*. Eizenga (1987) examined 44 tall fescue plants derived from anther-panicle cultures for GOT, MDH, PGI, ACP and 6-phosphogluconate dehydrogenase (6-PGD) and found different patterns for MDH and PGI. However, since the plants were not derived from the same callus, the different zymograms observed could be due to different origins of the regenerated plants. Dahleen and Eizenga (1990) compared isozyme banding patterns of PGI, GOT, 6-PGD, ACP and alcohol dehydrogenase (ADH) for 37 callus-derived tall fescue plants. Although variation was observed for all enzymes tested, only certain changes could be attributed to tissue culture because of possible heterogeneity of the explants used. Isozyme comparisons in a group of four tall fescue regenerants derived from the same embryo revealed differences in their *Got 3* and *Acp 1* loci. In addition, the PGI banding pattern indicated that all four regenerants had lost a *Pgi 2* allele found in the parent plant. This loss could have resulted from either tissue culture or genetic segregation (Dahleen and Eizenga 1990). In a larger scale study involving 174 tall fescue somaclones, variant alleles that were not present in the parents were identified in the somaclones for *Acp 1*, *Adh 1*, *Got 3* and *Pgi 2* (Eizenga and Cornelius 1991).

The most significant frequency differences were found in *Pgi 2* and *Pgd 1* phenotypes (Eizenga and Cornelius 1991). By grouping the regenerants into different categories, somaclones regenerated from longer term callus cultures were found to have a higher incidence of the less frequent, variant isozyme phenotypes than those regenerated earlier. No direct correlation was found between tissue culture-induced chromosome abnormalities and variant isozyme phenotypes (Eizenga and Cornelius 1991). However, since the somaclones analyzed were regenerated from different genotypes and at different time points, results from this study must be interpreted with caution. Studies using isozyme patterns in the analysis of somaclonal variation in *Festuca* should ideally be based on plants regenerated from single genotypes.

Humphreys and Dalton (1991) examined tall fescue plants regenerated from single genotype-derived cell suspensions and corresponding protoplasts for their *Pgi 2* phenotype. Special attention was paid to the *Pgi 2* locus since it is highly polymorphic in tall fescue (Humphreys 1989) and allows for the selection of individual genotypes with chromosomes labelled with different alleles (Humphreys and Dalton 1991). Among 180 cell suspension-derived regenerants, 11% showed different patterns in which both loss of alleles and the appearance of novel alleles were observed. All the protoplast-derived plants had the same *Pgi 2* phenotype as the donor plant but were hyperpolyploid, with chromosome numbers ranging between $2n = 60$ and $2n = 84$. This could account for their phenotypic stability in relation to the *Pgi 2* locus, since the loss or malfunction of alleles would be masked by genetic duplication in the hyperpolyploid regenerants (Humphreys and Dalton 1991). While analyzing another set of 76 tall fescue plants regenerated from single genotype-derived callus and cell suspensions, García et al. (1994) observed that up to 12.5% of the regenerants had deletions of *Pgi 2* alleles. This indicated that callus culture could also lead to high *Pgi 2* instability. A higher rate (17%) of *Pgi 2* allele deletions has been reported for plants regenerated from a cell suspension of pentaploid *L. multiflorum* × *F. arundinacea* hybrids (Humphreys and Dalton 1991) in which deletions were not always related to chromosome loss. However, direct relationship was observed between the frequency of aberrant *Pgi 2* phenotypes in regenerated plants and the age of the cell suspension from which they were derived (Humphreys and Dalton 1992).

Another application of isozyme banding patterns is the identification of sexual and somatic hybrids in grasses. Arcioni and Mariotti (1983) evaluated selfing and interspecific hybridization in *L. perenne* and *L. multiflorum* by using *Pgi 2* as an isozyme marker. This allowed not only for the determination of interspecific hybridization using either *L. perenne* or *L. multiflorum* as a seed parent, but also for the evaluation of self- and cross-fertilization of both parents (Arcioni and Mariotti 1983). Takamizo et al. (1991) used esterase patterns as one of the methods for the characterization of intergeneric somatic hybrid plants between *F. arundinacea* and *L. multiflorum*. Patterns of leaf esterases showed three bands unique to Italian ryegrass and three bands diagnostic of tall fescue. Esterase profiles for the somatic *Festulolium* hybrids revealed the presence of one *F. arundinacea*-specific and one *L. multiflorum*-specific band in addition to two bands common to both parents (Takamizo et al. 1991).

Genetic studies of PGI, indophenol oxidase (IPO) and peroxidase (PER) enzyme systems in Italian ryegrass showed that electrophoretically detectable variation in each system is governed by a single Mendelian locus with co-dominant alleles and that the three loci are inherited independently of each other (Polans and Allard 1985). The segregation of isozyme markers has also been examined for a set of microspore-derived progenies of *L. perenne* (Hayward et al. 1990). The range of variability found amongst the androgenetic progeny of perennial ryegrass does not conform with Mendelian expectation. The patterns of disturbed segregation for the alleles at *Pgi 2* and *Got 2* loci and the comparisons of green with albino plants indicated that the genetic component was responsible, possibly due to the linkage of a specific regeneration ability locus with the marker gene (Hayward et al. 1990).

The potential for using isozyme phenotypes as alternative selection criteria depends on the linkage of agronomic traits with marker loci. In order to establish whether any quantitative traits might be associated with one or more isozyme loci, Hayward and McAdam (1988) created a number of populations which were homozygous for differing alleles at the loci *Pgi 2*, *Got 2*, *Got 3* and *Acp 1* by selecting within three perennial ryegrass cultivars. Field assessment of the progenies of these selected populations revealed variation in yield and timing of inflorescence emergence. Differences in yield were associated with homozygosity for specific alleles at the *Pgi 2* locus, and it was suggested that the *Pgi 2* locus is possibly linked to a quantitative trait locus controlling yield (Hayward and McAdam 1988). A further study involving F_2 populations showed that there was a consistent association between water-soluble carbohydrate content and the genotype at the *Pgi 2* locus in perennial ryegrass, indicating that enzymes coded by the *Pgi 2* locus concerned with carbohydrate metabolism may eventually influence plant growth (Humphreys 1992). Furthermore, it was expected that selection for soluble carbohydrate content on the basis of the isozyme phenotype would likely be as efficient as family selection with regard to selection response (Humphreys 1992). Thus, the *Pgi 2* region of the genome of perennial ryegrass is of considerable importance and deserves further attention.

Allelic variation at different polymorphic enzyme loci has been used to study population structures of perennial ryegrass and red fescue (Livesey and Norrington-Davies 1991; Charmet et al. 1993; Balfourier and Charmet 1994) as well as genetic relationships between different *Lolium* species (Charmet and Balfourier 1994a). Some of the isozyme allele frequencies of *L. perenne* and other *Lolium* species from a survey of electrophoretic variation in a collection of natural populations of perennial ryegrass are illustrated in Fig. 9.1. These results confirmed that the population genetic statistics of perennial ryegrass fall into the range of values observed in other species of similar life

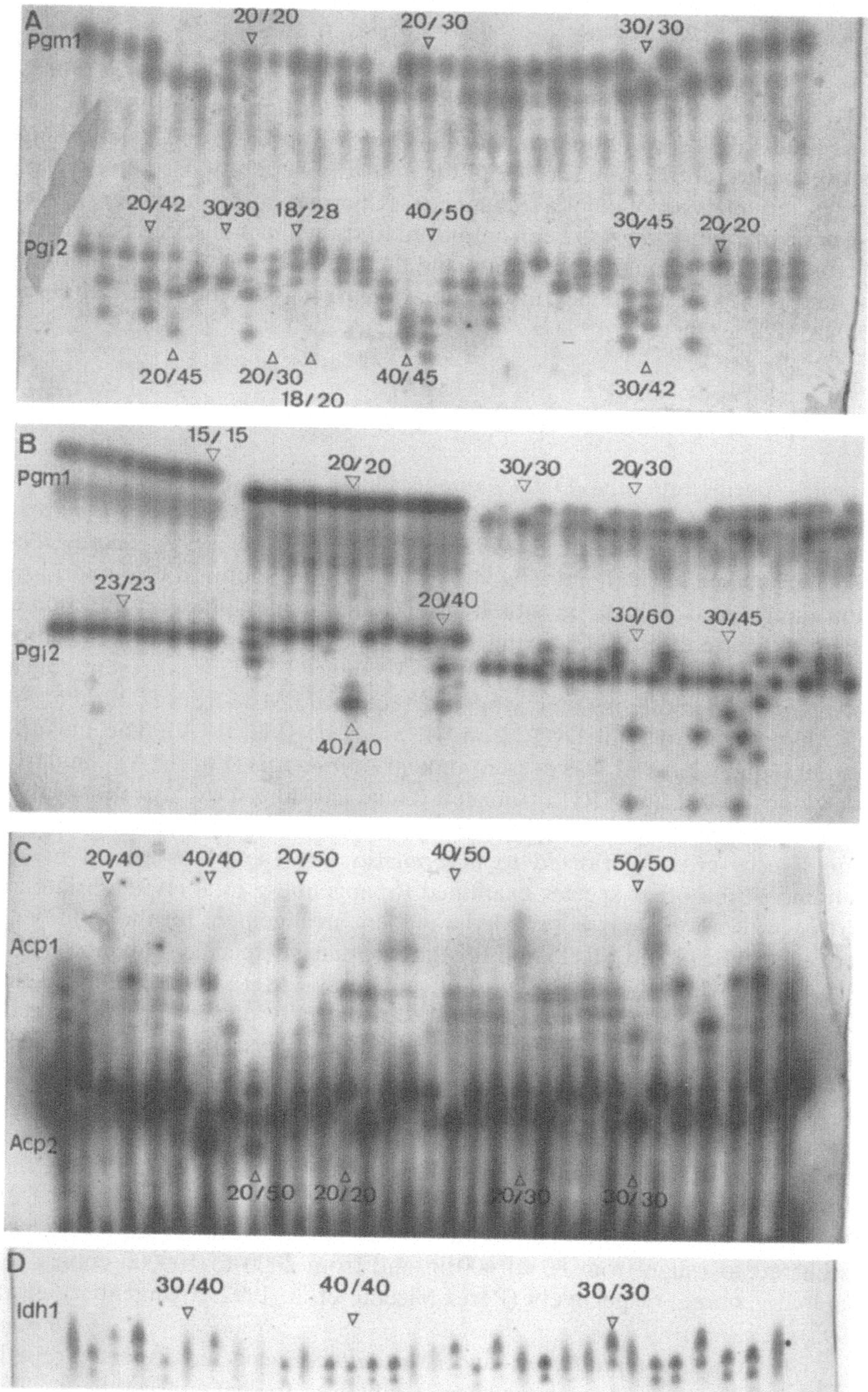

Fig. 9.1A–D. Isozyme allele frequencies of *L. perenne* and other *Lolium* species. **A** Phosphoglucomutase (*Pgm 1*) and phosphoglucose isomerase (*Pgi 2*) of *Lolium perenne* from France (*lanes 1–32*). **B** *Pgm 1* and *Pgi 2* of *L. persicum* (*lanes 1–10*), *L. canariense* (*lanes 11–22*) and *L. perenne* from Spain (*lanes 23–45*). **C** Acid phosphatases (*Acp 1* and *Acp 2*) of *L. perenne* (*lanes 1–32*). **D** Isocitric dehydrogenase (*Idh 1*) of *L. perenne* (*lanes 1–32*). (Courtesy of G. Charmet, INRA, Station d'Amélioration des Plantes, France)

cycle characteristics (Charmet et al. 1993). Clear differences in the frequency of certain alleles were found between *Lolium* species, and relationships between different *Lolium* species and *F. pratensis* were proposed on the basis of cluster analysis of allele frequencies (Charmet and Balfourier 1994a). Isozyme markers have also been employed to study the influence of ecological factors on population differentiation and the conservation of allelic multiplicity and genotypic frequency in perennial ryegrass (Charmet and Balfourier 1994b; Balfourier et al. 1994).

9.2.2 Species-Specific Repetitive DNA Sequences

Species-specific repetitive DNA sequences have been isolated from different grass species: *L. multiflorum*, *F. arundinacea*, *F. rubra* and *Alopecurus pratensis* (Pérez-Vicente et al. 1992; Spangenberg et al. 1995). These DNA sequences have been characterized by dot blot and Southern blot hybridization analyses as well as in situ hybridization to metaphase chromosomes (Pérez-Vicente et al. 1992; Spangenberg et al. 1995). Dot blot hybridization analysis revealed that the Italian ryegrass sequence LMH2 was present only in *L. multiflorum* and *L. perenne*, while the sequence LMB4 from *L. multiflorum* also hybridized to total DNA from *A. pratensis* (Fig. 9.2A). The meadow foxtail sequence APE1 was present only in *A. pratensis* (Fig. 9.2A). Similarly, the clones FRH2 and FRH5 from red fescue hybridized exclusively to DNA from *F. rubra* (Fig. 9.2A), while the sequence FAH1 cloned from *F. arundinacea* cross-hybridized to *F. pratensis* DNA but was absent in the genome of the other species examined (Spangenberg et al. 1995). Southern blot hybridization analysis revealed a tandem arrangement for the red fescue sequences FRH2 and FRH5 and for the *A. pratensis*-specific sequence APE1 (Fig. 9.2B), while the *L. multiflorum* sequences LMH2 and LMB4 and the *F. arundinacea* sequence FAH1 showed a partially dispersed arrangement (Pérez-Vicente et al. 1992). In situ hybridization analysis revealed that FRH2 and FRH5 only hybridized to 6 and 2 chromosomes of red fescue, respectively (Fig. 9.2C), while APE1 hybridized to 24 meadow foxtail chromosomes (Spangenberg et al. 1995). LMH2 and LMB4 hybridized to all Italian ryegrass chromosomes (Fig. 9.2D), and hybridization to only some of the tall fescue chromosomes was observed for FAH1 (Fig. 9.2E; Pérez-Vicente et al. 1992). Insert size and copy number of the highly species-specific sequences was estimated to range from 80 to 400 bp and from 2000 to 100000 copies per haploid genome, respectively (Pérez-Vicente et al. 1992; Spangenberg et al. 1995).

The cloned *L. multiflorum*- and *F. arundinacea*-specific dispersed repetitive DNA sequences have been successfully used to characterize the nuclear composition of symmetric and asymmetric intergeneric *F. arundinacea* (+) *L. multiflorum* somatic hybrid plants (Chap. 6). The species-specific repetitive DNA sequences from red fescue and Italian ryegrass have proven useful tools

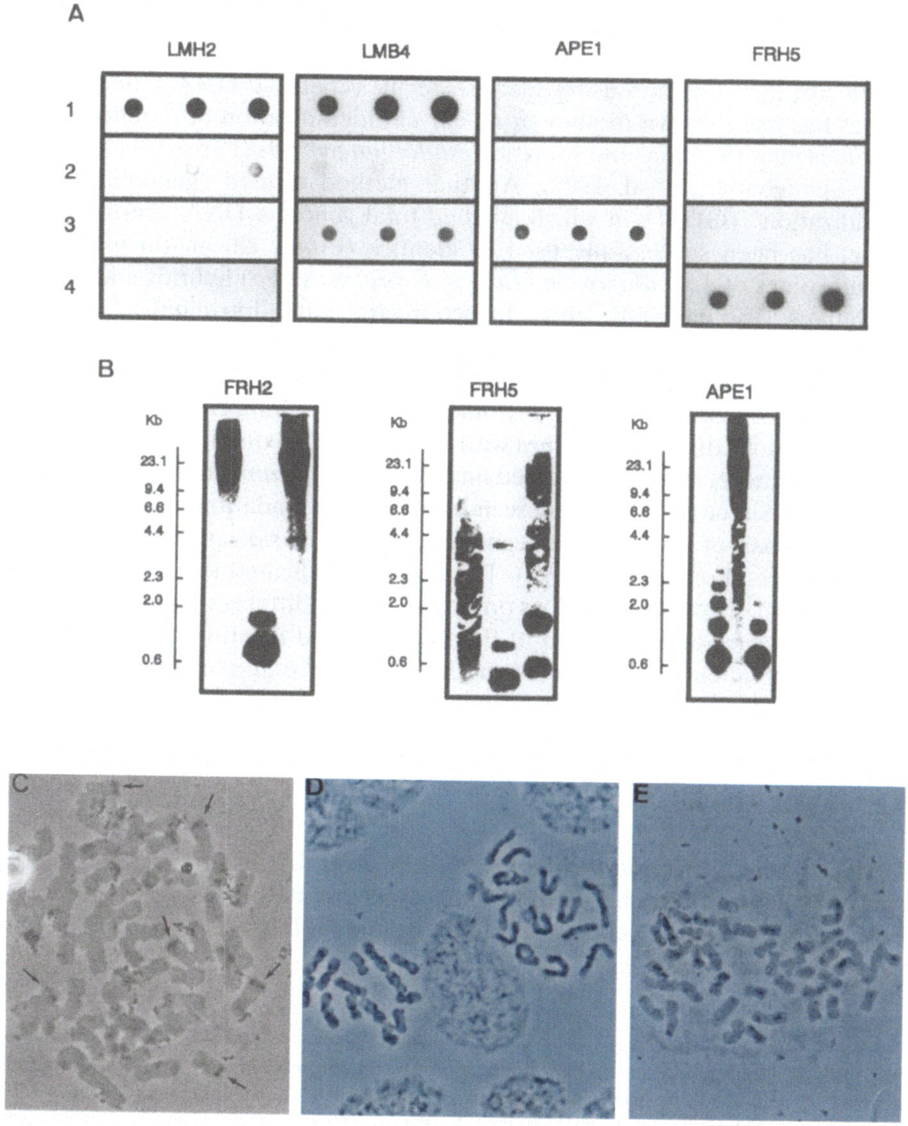

Fig. 9.2A–E. Isolation and characterization of species-specific repetitive DNA sequences in *Festuca*, *Lolium* and *Alopecurus*. **A** Test for species specificity of repetitive DNA sequences isolated from *L. multiflorum* (*LMH2* and *LMB4*), *A. pratensis* (*APE1*) and *F. rubra* (*FRH5*). Labelled probes were hybridized to dot blots containing (from *left* to *right*) 0.5, 1 and 2 μg of *1 L. multiflorum*; *2 L. perenne*; *3 A. pratensis*; and *4 F. rubra* DNA. **B** Southern blot hybridization patterns from *F. rubra* and *A. pratensis* DNA probed with repetitive DNA sequences *FRH2*, *FRH5* and *APE1*, respectively. In all cases, 7 μg of total DNA digested with *Bam*HI, *Hind*III or *Eco*RI (*lanes* from *left* to *right*) were loaded. **C–E** In situ hybridization analysis of repetitive DNA sequences from: **C** *F. rubra* (FRH2), **D** *L. multiflorum* (LMH2), and **E** *F. arundinacea* (FAH1) hybridized to metaphase chromosomes of red fescue, Italian ryegrass and tall fescue, respectively. *Arrows* indicate hybridization signals on red fescue chromosomes (**C**). (Spangenberg et al. 1995)

for the analysis of the genetic make-up of *F. rubra* (+) *L. perenne* somatic hybrids (Spangenberg et al. 1995).

In situ hybridization using species-specific repetitive DNA sequences as probes has been shown to allow for clear-cut identification of parental chromosomes in both sexual and somatic *Festulolium* hybrids (Pérez-Vicente et al. 1992; Spangenberg et al. 1994). Another method termed "genomic in situ hybridization" (GISH), in which labelled total genomic DNA is used as the probe, has been successfully used to identify *Festuca* chromatin in mitotic chromosomes of *L. multiflorum* (*Lm*) × *F. pratensis* (*Fp*) hybrids and hybrid derivatives (Thomas et al. 1994). In two inverse autoallotriploids, *LmLmFp* and *LmFpFp*, GISH was able to discriminate between the *Lolium* and *Festuca* chromosomes (Fig. 9.3A, B). In a third triploid hybrid produced by crossing an amphiploid of *L. multiflorum* × *F. pratensis* with *L. multiflorum*, the technique was used to identify chromosomes with interspecific recombination (Fig. 9.3C, D). Furthermore, in an introgressed line of *L. multiflorum* which was homozygous for the recessive *sid* (senescence-induced degradation) allele from *F. pratensis*, a pair of chromosome segments carrying the *sid* gene were identified (Fig. 9.3E, F; Thomas et al. 1994). These results indicated that in some cases the direct use of genomic DNA as probes allows for differentiation of chromosomes of species and detection of the presence and location of introgressed segments of alien chromosomes in the *Festuca-Lolium* complex, without species-specific repetitive DNA sequences having to be isolated.

9.2.3 RFLP Markers

Within the *Festuca-Lolium* complex, RFLPs were first used to assess genetic variation among different *F. arundinacea* accessions by testing single copy probes isolated from a genomic library of hexaploid tall fescue (Xu and Sleper

→

Fig. 9.3A–F. Fluorescence in situ hybridization of root-tip chromosome preparations using total genomic *Festuca* DNA as probe labelled with **A** biotin and **B** digoxigenin, each detected with fluorescein, and **D, F** labelled with rhodamine. **A** *LmLmFp* (triploid hybrids obtained by crossing autotetraploid *L. multiflorum* with diploid *F. pratensis*); the seven *Festuca* chromosomes show hybridization (*yellow*), and the *Lolium* chromosomes are counterstained *red* with propidium iodide. **B** *LmFpFp* (triploid hybrids obtained by crossing autotetraploid *F. pratensis* with diploid *L. multiflorum*); the 14 *Festuca* chromosomes are hybridized (*bright yellow*), and the *Lolium* chromosomes are *dull*. **C** *Lm* (*LmFp Elmet*) stained with 4,6-diamidino-2-phenylindole (DAPI). **D** The same cell as **C** showing sites of hybridization to *Festuca* chromosomes and chromosome segments (*bright red*), while the *Lolium* chromosomes and chromosome segments remain *dull*. **E** An incomplete cell of the *L. multiflorum* introgression line *sid-sid* stained with DAPI. **F** The same cell as in **E**; segments of *Festuca* chromosomes show hybridization in the distal parts of the long arms of one pair of chromosomes (*arrows*) that carry the *sid* gene and in the centrometric region of another chromosome (*arrowhead*). *Bar* represents 10 μm. (Thomas et al. 1994; courtesy of H. M. Thomas, Institute of Grassland and Environmental Research, Aberystwyth, UK)

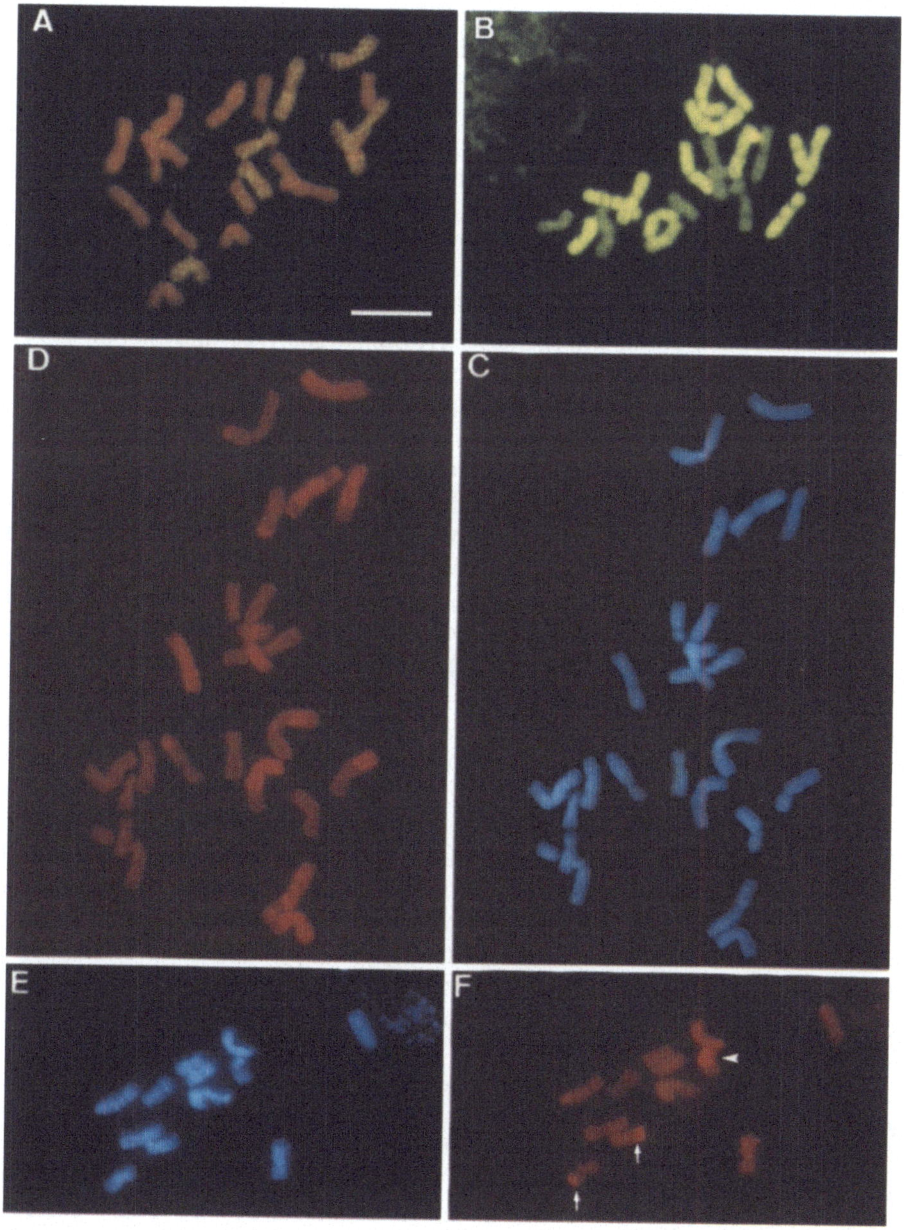

1991). Out of 174 clones evaluated from the library, 36 (21%) were found to be repetitive, and 138 (79%) were single or low copy. The single copy probes revealed a considerable extent of polymorphism among the tall fescue accessions tested (Xu and Sleper 1991). In this study hexaploids showed more polymorphism than tetraploids, which in turn exhibited more variation than diploid *Festuca* species. Among diploids, 30–34% of the polymorphism was detected by using three restriction enzymes, while in the tetraploid and hexaploid species, 41–43% and 67–72% of the polymorphism was detected with these restriction enzymes, respectively (Xu and Sleper 1991).

RFLPs have also been used to examine the phylogenetic relationship of tall fescue with six related species (Xu and Sleper 1994). Fourteen probes out of a total of 29 genomic clones used showed differential hybridization patterns, especially at the diploid and tetraploid levels. These genome-specific molecular markers may be useful for verifying hybrids and monitoring gene introgression in interspecific and intergeneric hybridizations (Xu and Sleper 1994). Dendrograms based on RFLP data reflected the phylogenetic relationships among these species and indicated that tall fescue has a close relationship with *F. pratensis* (diploid), *F. arundinacea* var. *glaucescens* (tetraploid) and *L. perenne* (diploid) and that *F. pratensis* and *L. perenne* have the closest degree of relationship (Xu and Sleper 1994). Furthermore, the RFLP probes from *F. arundinacea* have also been used to investigate genetic variations among and within tall fescue cultivars (Xu et al. 1994). The number of bands was significantly different among cultivars and was related to the genetic diversity within cultivars. The dendrogram from the corresponding cluster analysis reflected the known pedigree information of the tall fescue cultivars (Xu et al. 1994).

The genomic clones from tall fescue have also been used to detect RFLPs in *Lolium* (Xu et al. 1992; Chen et al. 1997). Two-thirds of the 39 heterologous probes tested from tall fescue cross-hybridized to *Lolium* DNA, and high levels of polymorphism were detected. It was estimated that an average 41% of the loci examined were homozygous, ranging from 21 to 58% among the five perennial ryegrass genotypes evaluated (Xu et al. 1992). In addition, tall fescue clones used in this study exhibited the same cross-hybridization behaviour with *Lolium* as was observed with meadow fescue, suggesting that the *Lolium* genome is closely related to that of *F. pratensis* (Xu et al. 1992). The genomic clones from tall fescue were successfully used for RFLP detection in studies on mechanisms of 2n pollen formation in diploid *L. perenne* (Chen et al. 1997).

Genetic uniformity of meadow fescue plants regenerated from single genotype-derived embryogenic cell suspensions and corresponding protoplasts was analyzed by a set of RFLP markers involving three mitochondrial gene-specific probes, one chloroplast gene-specific probe and one nuclear repetitive DNA sequence isolated from *F. arundinacea* (Vallés et al. 1993). No variation in either sets of plants was detected in the mitochondrial, plastidial or nuclear genomes. This is not surprising, since the same hybridization probes failed to detect polymorphisms among different genotypes within one cultivar and even in some cases among different cultivars of *F. pratensis*, although they did reveal RFLPs in different *Festuca* species (Vallés et al. 1993).

A number of heterologous mitochondrial probes, *cox1* (cytochrome c oxidase subunit 1), *cox2*, *cox3*, *atp6* (ATPase subunit 6) and *atpA* (ATPase

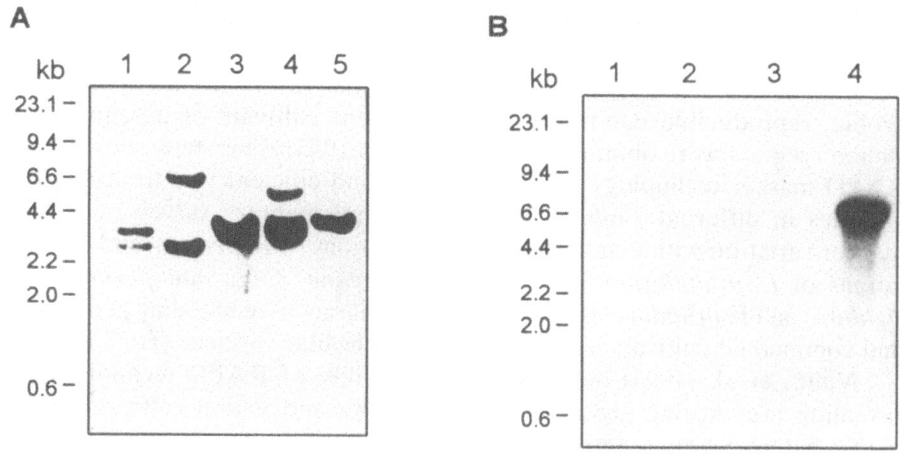

Fig. 9.4A,B. RFLP analysis of male fertile and cytoplasmic male-sterile (CMS) *Lolium* lines. **A** *Hin*dIII-digested genomic DNA hybridized with the mitochondrial DNA probe *cox1*: *1 L. multiflorum* cv. Axis; *2 L. multiflorum* CMS line WS17; *3 L. perenne* cv. Citadel; *4 L. perenne* CMS line Lp9A; *5 L. perenne* CMS line 7B290. **B** *Hin*dIII-digested genomic DNA hybridized with the diagnostic probe pCMS45 isolated from the CMS line 7B290: *1 L. perenne* cv. Citadel; *2 L. perenne* inbred line; *3 L. perenne* CMS line Lp9A; *4 L. perenne* CMS line 7B290

subunit A), have been used to characterize the mitochondrial DNA organization of male-sterile and fertile perennial ryegrass plants (Rouwendal et al. 1992). Southern hybridization analyses using these heterologous probes allowed clear discrimination between the cytoplasmic types (Fig. 9.4A). Furthermore, the transcriptional patterns of *atp6* and *cox1* genes distinguished between male-sterile and fertile cytoplasms (Rouwendal et al. 1992).

A diagnostic probe was developed for a male-sterile cytoplasm in *L. perenne* (Kiang et al. 1993). A 4.5 kb restriction fragment (pCMS45) was identified and cloned from the mtDNA of the cytoplasmic male-sterile (CMS) line. This fragment failed to hybridize to sequences in the mtDNA of fertile lines and is thus capable of unambiguously distinguishing between fertile and CMS cytoplasms (Fig. 9.4B) as a diagnostic probe for simple identification of young vegetative plants carrying the male-sterile cytoplasm (Kiang et al. 1993).

9.2.4 RAPD Markers

RAPD markers have been applied to cultivar identification and evaluation of genetic stability of tissue culture-derived plants in forage grasses. De Loose et al. (1993) compared RAPD patterns from 22 ryegrass cultivars (2 *L. multiflorum* ssp. westerwoldicum, 6 *L. multiflorum* and 14 *L. perenne* cultivars). Out of a total of 60 primers used, 7 gave rise to clearly different banding profiles. For each of these 7 primers evaluated, a multiple band profile or fingerprint was produced which comprised one to five major bands and a varying number of minor bands. By computer analysis of each multiple RAPD profile, reproducible dendrograms for different cultivars of perennial and Italian ryegrass were obtained (De Loose et al. 1993). This study showed that RAPD marker technology provides a quick and efficient way to distinguish varieties in different *Lolium* species. A complementary survey of RAPD marker variation within and among 11 populations of *L. perenne* and 6 populations of *L. multiflorum*, *L. rigidum*, *L. perenne* × (*L. multiflorum* × *L. rigidum*) and *F. arundinacea* was reported to characterize breeding germplasm and commercial cultivars using analysis of molecular variance (Huff 1997).

Vallés et al. (1993) demonstrated the utility of RAPD technology for revealing pre-existing genetic variation among and within cultivars in the highly heterozygous outcrossing species *F. pratensis*. An analysis based on RAPD markers using 18 different primers allowed for discrimination between different *Festuca* species and even between different genotypes within one meadow fescue cultivar (Fig. 9.5A; Vallés et al. 1993). When analyzing independent plants regenerated from a single genotype-derived cell suspension and corresponding protoplasts, no newly generated variation among the regenerants was revealed (Fig. 9.5A), suggesting that the regenerated plants were genetically stable at the loci screened with the molecular markers (Vallés et al. 1993).

RAPD markers have also proven suitable for detecting pre-existing variations in Italian ryegrass (Wang et al. 1993b). They allowed discrimination of different *Lolium* species and different cultivars and genotypes in *L. multiflorum* (Fig. 9.5B–D). Most of the plants regenerated from protoplasts isolated from a single genotype-derived cell suspension of *L. multiflorum* included in the RAPD analysis showed identical patterns for most of the

Fig. 9.5A–D. Analysis of genetic uniformity of *Festuca* and *Lolium* plants regenerated from cell suspensions and protoplasts screened by RAPD markers. **A** Screening by RAPDs generated with primer OPU-8: *1* different *Festuca* species (*F. pratensis*, *F. arundinacea*); *2* independent cell suspension-derived plants of a single genotype of *F. pratensis* cv. Barmondo; *3* independent *F. pratensis* cv. Barmondo genotypes; *4* independent protoplast-derived plants regenerated from one cell suspension of *F. pratensis* cv. Barmondo. **B** Screening by RAPDs generated with primer OPU-3: *1* independent protoplast-derived plants regenerated from a single suspension culture of *L. multiflorum* cv. Andy, showing identical patterns; *2* plants from different cultivars in *L. multiflorum* (cvv. Andy, Caramba, Axis, Lipo, Limella and Fedo). **C** Screening by RAPDs generated with primer OPU-16: *1* independent protoplast-derived plants regenerated from one cell suspension of *L. multiflorum* cv. Andy, showing identical patterns; *2* independent genotypes from *L. multiflorum* cv. Andy. **D** Screening by RAPDs generated with primer OPU-8: *1* independent protoplast-derived plants regenerated from one cell suspension of *L. multiflorum* cv. Andy, some showing different patterns; *2* plants from different cultivars in *L. multiflorum* (cvv. Andy, Caramba, Axis, Lipo, Limella and Fedo); *3* independent genotypes from *L. multiflorum* cv. Andy; *4* different *Lolium* species (*L. multiflorum*, *L. perenne* and *L.* × *boucheanum*); **M** 1kb-ladder size markers. (Vallés et al. 1993; Wang et al. 1993b)

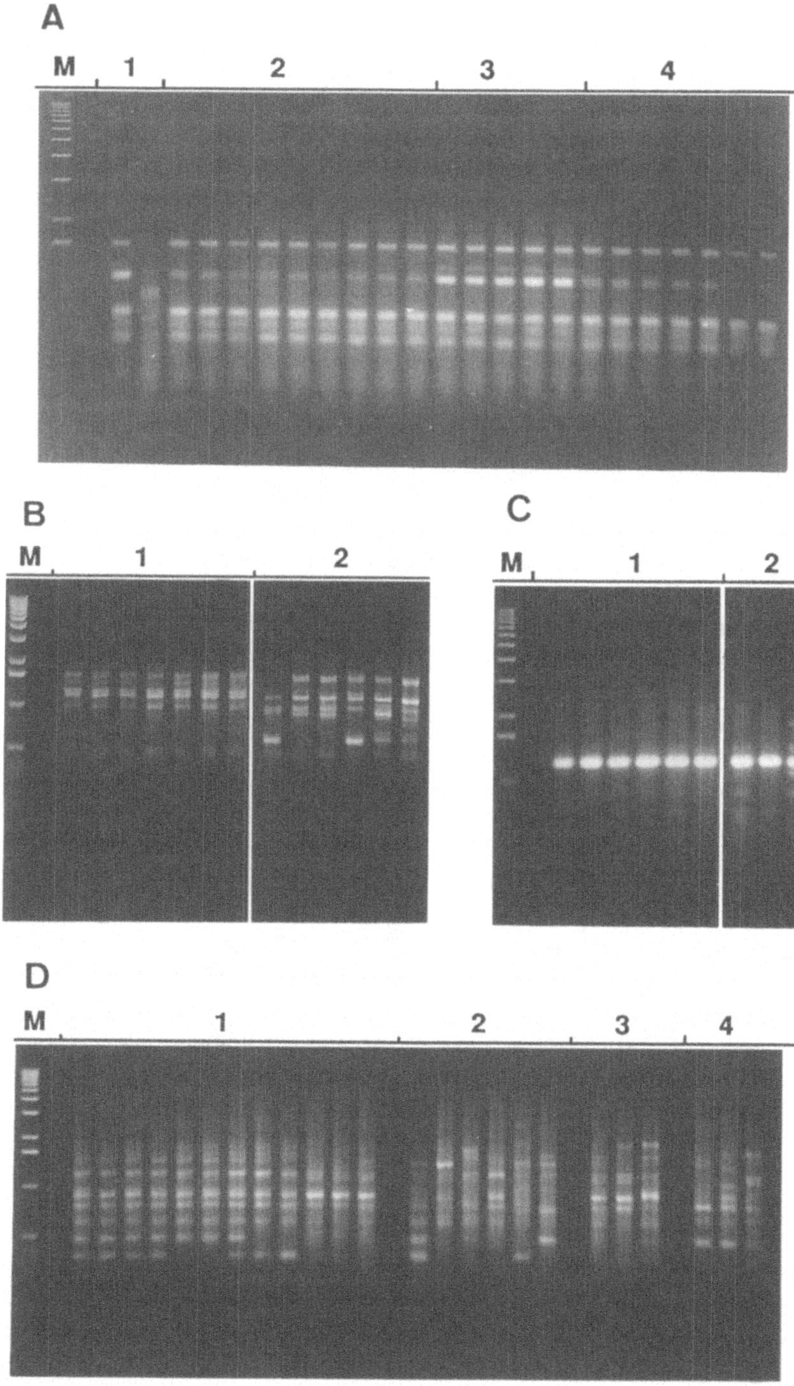
A
M
1
2
3
4
B
M
1
2
C
M
1
2
D
M
1
2
3
4

primers used (Fig. 9.5B, C). Limited newly induced genetic variation among independent protoplast-derived Italian ryegrass plants was detected at the loci screened for 3 out of 19 primers tested (Fig. 9.5D; Wang et al. 1993b).

Stammers et al. (1995) employed the RAPD technique to investigate phylogenetic relationships between species of the genera *Lolium* and *Festuca*. The degree of band sharing was used to evaluate genetic distances between species and to construct a phylogenetic tree which showed good overall agreement with classical taxonomy, yet with a number of novel insights (Stammers et al. 1995). The *Lolium* species were all grouped on the phylogenetic tree, with *L. multiflorum* and *L. perenne* clustered together and the annual species (*L. temulentum*, *L. rigidum* and *L. loliaceum*) placed at a greater distance from this grouping. One of the striking features of the phylogenetic tree obtained was the placement of *F. pratensis* among the members of the *Lolium* genus, suggesting a close relationship between *F. pratensis* and the outbreeding *Lolium* species. This observation is in agreement with results obtained from RFLP analyses reported by Xu et al. (1992) and Xu and Sleper (1994). Consistent results were also obtained from the phylogenetic interpretation of the use of RAPD and RFLP markers, as well as restriction site analysis of non-coding chloroplast DNA and internal transcribed spacer (ITS) sequences of nuclear rDNA in the *Festuca-Lolium* complex (Charmet et al. 1997). The physical mapping of 18S-5.8S-26S and 5S rRNA genes in *F. arundinacea*, *F. pratensis*, *F. gigantea* and *F. mairei* by GISH and comparison of number and position of rDNA sites confirmed earlier phylogenetic studies of these diploid, tetraploid and hexaploid *Festuca* species and provided evidences for the occurrence of some structural rearrangements and loss of rDNA sites during polyploidization (Thomas et al. 1997).

A first genetic map of *Lolium* has been constructed after analyzing the segregation of 101 loci in 89 plants from an (*L. perenne* × *L. multiflorum*) × doubled-haploid *L. perenne* mapping family (Hayward et al. 1994). The markers employed were isozymes, RFLPs and RAPDs. Thirteen linkage groups were identified with 19 of the markers being unlinked. After omitting some RAPD markers showing disturbed segregations, the total map length was of the order of 750 cM. When related to the cytological data for *Lolium*, this preliminary map is still rather long, especially considering that linkage groups will eventually be joined together (Hayward et al. 1994).

9.3 Molecular Markers in Other Grasses

Isozyme banding patterns have been examined for cultivar identification in the apomictic species *Poa pratensis*. Wehner et al. (1976) first succeeded in identifying 11 out of 15 Kentucky bluegrass cultivars on the basis of peroxidase banding patterns. With the exception of 2 lines, Spoor and Hay (1979) were able to identify 14 lines of Kentucky bluegrass using esterase banding patterns alone or in conjunction with peroxidase zymograms. Wu et al. (1984) examined 24 Kentucky bluegrass cultivars for isozyme variations in esterase, phosphoglucomutase (PGM), phosphoglucoisomerase (PGI) and glutamate-oxaloacetate transaminase (GOT). It was demonstrated that the combination of PGM and esterase isozyme systems provided a high degree of resolution in distinguishing among Kentucky bluegrass cultivars, where 22 out of 24 cultivars showed distinct differences for one or both enzymes (Wu et al. 1984). In addition, variation within seed lots was also found, indicating that sexual segregation has occurred in some cases (Wu et al. 1984). The usefulness of PGM and esterase for cultivar identification was confirmed by Weeden and Emmo (1985). Furthermore, the authors found that glucose phosphate isomerase (GPI) and triose phosphate isomerase (TPI) also provided excellent means for the identification of certain Kentucky bluegrass cultivars (Weeden and Emmo 1985).

For the cross-pollinated *Agrostis* species, PGI, TPI, GOT and peroxidase (PER) banding patterns provided useful information allowing the characterization of bentgrass species and cultivars (Yamamoto and Duich 1994). The locus *Pgi 2* showed the most distinctive polymorphism. Except for 2 cultivars each from *A. capillaris* and *A. castellana*, the *Pgi 2* banding patterns allowed for the identification of 24 out of 26 tested cultivars which included 12 *A. palustris*, 10 *A. capillaris*, 1 *A. castellana*, 2 *A. gigantea* and 1 *A. canina* cultivars (Yamamoto and Duich 1994).

Somaclonal variation of tissue culture-derived Kentucky bluegrass plants has been evaluated by esterase and PGM patterns (Wu and Jampates 1986). Since all 57 callus-derived plants analyzed showed isozyme patterns identical to the parent plants, no somaclonal variation was found (Wu and Jampates 1986). Similar results were obtained when analyzing napier grass plants regenerated from callus cultures using 14 isozyme systems, but no qualitative variation was found at any of the loci examined (Shenoy and Vasil 1992). Furthermore, in this study, no RFLPs were detected in the mitochondrial, plastidial or nuclear genomes of the regenerated napier grass plants (Shenoy and Vasil 1992).

An RFLP-based genetic map of *Pennisetum glaucum* has been generated (Liu et al. 1994). The analysis of a sample of diverse pearl millet genotypes with 200 genomic DNA probes revealed this species to be extremely polymorphic. Eighty-five percent of the clones showed polymorphic hybridization patterns when digested with just two enzymes, and the average pairwise polymorphism between all the probe-enzyme combinations was 56%. The total length of the pearl millet RFLP map, which comprised seven linkage groups, was 303 cM. The average distance between loci was 2 cM (Liu et al. 1994).

RFLPs have been combined with RAPDs to analyze the apomictic mode of reproduction in *Pennisetum* (Ozias-Akins et al. 1993), where RAPDs proved more efficient in early identification of different phenotypes. The molecular analysis of advanced backcross individuals has demonstrated co-inheritance of apomictic reproduction and two molecular markers, one an RFLP marker and the other a RAPD marker (Ozias-Akins et al. 1993).

RAPD markers, especially when combined with flow cytometry, allowed the determination of the genetic origins of aberrant progeny which are genetically different from the maternal genotype in facultative apomictic Kentucky bluegrass (Huff and Bara 1993; Barcaccia et al. 1997). This ability has powerful applications in detecting hybridization events during early stages of a breeding program and in genetic studies analyzing variation in apomixis pathway components and the frequency of different genetic origins (Huff and Bara 1993; Barcaccia et al. 1997). Isozyme and RAPD markers have also been used for the early identification of non-maternal plants in progenies of facultative apomictic Kentucky bluegrass (Mazzucato et al. 1995). The RAPD method proved the most precise screening technique when compared with isozyme and morphological analyses (Mazzucato et al. 1995).

In outcrossing buffalograss (*Buchloë dactyloides*), RAPD markers have been shown to provide a powerful tool for the investigation of genetic variation in natural populations (Huff et al. 1993).

DNA amplification fingerprinting (DAF), another PCR-based technique similar to the RAPD technique but based on shorter primers and the analysis of amplification products resolved in silver-stained polyacrylamide gels, has proven successful in evaluating genetic diversity and distinguishing cultivars and hybrids in centipedegrass (*Eremochloa ophiuroides*; Weaver et al. 1995) and bermudagrass (*Cynodon* spp; Caetano-Anollés et al. 1995; Ho et al. 1997).

9.4 Summary and Conclusions

The identification of polymorphic loci in forage and turf grasses has come about with the advent of isozymes and DNA-based markers.

Isozyme polymorphism is well known in grasses and has a relatively long history. Isozymes as near-neutral genetic markers have provided a powerful means for cultivar discrimination in *Lolium*, *Poa* and *Agrostis*. They have also been used for hybrid identification, evaluation of somaclonal variation and population structure studies in different forage and turf grasses.

The development of DNA-based markers and genetic maps in forage and turf grasses has been relatively slow as compared to some major cereal crops. Nevertheless, RAPDs and DAF have been used for cultivar identification, evaluation of genetic stability of tissue culture-derived plants, and the analysis of phylogenetic relationships within the *Festuca-Lolium* complex. RFLP markers have revealed genome divergence and allowed for the detection of high levels of polymorphism in *Festuca* and *Lolium*.

Progress has been made in the construction of linkage maps of *L. perenne* and *P. glaucum*, which have provided a basis for determining possible markers linked to loci controlling traits of agronomic importance. The use of markers is likely to have its greatest impact in the manipulation of quantitative characters. Future efforts at development and deployment of user-friendly molecular marker technology in outcrossing grasses are expected to be focused on: (1) the development and mapping of highly informative codominant – preferably PCR-based – genetic markers, such as simple sequence repeats or microsatellites, to provide a framework genetic map for use in trait linkage studies; (2) the comparative gene mapping intended to benefit from spin-offs of the far more advanced mapping work in cereals; and (3) the identification and location of quantitative trait loci (QTLs) for a wide range of key target characters, by one or more linked markers, to provide the capability for marker assisted selection. The generation of enhanced genetic maps will be essential for marker-assisted selection since they allow for more precisely identifying superior individuals and hence enhancing selection efficiency. Concerted international efforts focused on a key grass species, such as *L. perenne*, will accelerate the development of an enhanced genetic map underpinning the deployment of genetic markers for marker-assisted selection.

References

Arcioni S, Mariotti D (1983) Selfing and interspecific hybridization in *Lolium perenne* L. and *Lolium multiflorum* Lam. evaluated by phosphoglucoisomerase as isozyme marker. Euphytica 32: 33–40

Balfourier F, Charmet G (1994) Geographical patterns of isozyme variation in Mediterranean populations of perennial ryegrass. Heredity 72: 55–63

Balfourier F, Charmet G, Grand-Ravel C (1994) Conservation of allelic multiplicity and genotypic frequency by pooling wild populations of perennial ryegrass. Heredity 73: 386–396

Baird E, Cooper-Bland S, Waugh R, DeMaine M, Powell W (1992) Molecular characterisation of inter- and intra-specific somatic hybrids of potato using randomly amplified polymorphic DNA (RAPD) markers. Mol Gen Genet 233: 469–475

Barcaccia G, Mazzucato A, Belardinelli A, Pezzotti M, Lucretti S, Falcinelli M (1997) Inheritance of parental genomes in progenies of *Poa pratensis* L. from sexual and apomictic genotypes as assessed by RAPD markers and flow cytometry. Theor Appl Genet 95: 516–524

Beckmann JS, Soller M (1986) Restriction fragment length polymorphism and genetic improvement of agricultural species. Euphytica 35: 111–124

Brown AHD (1978) Isozymes, plant population genetic structure and genetic conservation. Theor Appl Genet 52: 145–157

Burr B (1994) Some concepts and new methods for molecular mapping in plants. In: Phillips RL, Vasil IK (eds) DNA-based markers in plants. Kluwer Academic, Dordrecht, pp 1–7

Caetano-Anollés G, Callahan LM, Williams PE, Weaver KR, Gresshoff PM (1995) DNA amplification fingerprinting analysis of bermudagrass (*Cynodon*): genetic relationships between species and interspecific crosses. Theor Appl Genet 91: 228–235

Charmet G, Balfourier F (1994a) Isozyme variation and species relationships in the genus *Lolium* L. (ryegrasses, Graminaceae). Theor Appl Genet 87: 641–649

Charmet G, Balfourier F (1994b) Influence of ecological factors on population differentiation in perennial ryegrass (*Lolium perenne* L.). Genet Resourc Crop Evol 41: 175–184

Charmet G, Balfourier F, Ravel C (1993) Isozyme polymorphism and geographic differentiation in a collection of French perennial ryegrass populations. Genet Resourc Crop Evol 40: 77–89

Charmet G, Ravel C, Balfourier F (1997) Phylogenetic analysis in the *Festuca-Lolium* complex using molecular markers and ITS rDNA. Theor Appl Genet 94: 1038–1046

Chen C, Sleper DA, Chao S, Johal GS, West CP (1997) RFLP detection of 2n pollen formation by first and second division restitution in perennial ryegrass. Crop Sci 37: 76–80

Dahleen LS, Eizenga GC (1990) Meiotic and isozymic characterization of plants regenerated from euploid and selfed monosomic tall fescue embryos. Theor Appl Genet 79: 39–44

De Loose M, Vauterin L, Reuheul D, Van Bockstaele E (1993) Identification of ryegrass varieties using RAPD markers. Med Fac Landbow Univ Gent 58/4b: 2119–2121

Eizenga GC (1987) Cytogenetic and isozymic characterization of anther panicle culture derived tall fescue aneuploids. Euphytica 36: 175–179

Eizenga GC, Cornelius PL (1991) Comparison of the isozyme variation in tall fescue parents and their somaclones. Euphytica 51: 249–256

Evola SV, Burr FA, Burr B (1986) The suitability of restriction fragment length polymorphisms as genetic markers in maize. Theor Appl Genet 71: 765–771

García A, Dalton SJ, Humphreys MO (1994) Reproductive disturbances and phosphoglucoisomerase instability in *Festuca arundinacea* (tall fescue) plants regenerated from callus and cell suspension cultures. Heredity 73: 355–362

Gilliland TJ, Camlin MS, Wright CE (1982) Evaluation of phosphoglucoisomerase allozyme electrophoresis for the identification and registration of cultivars of perennial ryegrass (*Lolium perenne*). Seed Sci Technol 10: 415–430

Greneche M, Lallemand J, Michaud O (1991) Comparison of different enzyme loci as a means of distinguishing ryegrass varieties by electrophoresis. Seed Sci Technol 19: 147–158

Hayward MD, McAdam NJ (1977) Isozyme as a measure of distinctiveness and stability in cultivars of *Lolium perenne*. Z Pflanzenzüchtg 79: 59–68

Hayward MD, McAdam NJ (1988) The effect of isozyme selection on yield and flowering time in *Lolium perenne*. Plant Breed 101: 24–29

Hayward MD, Olesen A, Due IK, Jenkins R, Morris P (1990) Segregation of isozyme marker loci amongst androgenetic plants of *Lolium perenne* L. Plant Breed 104: 68–71

Hayward MD, McAdam NJ, Jones JG, Evans C, Evans GM, Forster JW, Ustin A, Hossain KG, Quader B, Stammers M, Will JK (1994) Genetic markers and the selection of quatitative traits in forage grass. Euphytica 77: 269–275

Helentjaris TG (1992) RFLP analysis for manipulating agronomic traits in plants. In: Stalker HT, Murphy JP (eds) Plant breeding in the 1990s. CAB International, Wallingford, pp 357–387

Hicks PH, Sleper DA, Randall DD, Crane CF (1982) Peroxidase isozyme differences in tall fescue cultivars and allopolyploid accessions. Euphytica 31: 175–181

Ho CY, McMaugh SJ, Wilton AN, McFarlane IJ, Mackinlay AG (1997) DNA amplification variation within cultivars of turf-type Couch grasses (*Cynodon* spp.). Plant Cell Rep 16: 797–801

Huff DR, Bara JM (1993) Determining genetic origins of aberrant progeny from facultative apomictic Kentucky bluegrass using a combination of flow cytometry and silver-stained RAPD markers. Theor Appl Genet 87: 201–208

Huff DR, Peakall R, Smouse PE (1993) RAPD variation within and among natural populations of outcrossing buffalograss [*Buchloë dactyloides* (Nutt.) Engelm.]. Theor Appl Genet 86: 927–934

Huff DR (1997) RAPD characterization of heterogeneous perennial ryegrass cultivars. Crop Sci 37: 557–564

Humphreys MW (1989) The controlled introgression of *Festuca arundinacea* genes into *Lolium multiflorum*. Euphytica 42: 105–116

Humphreys MO (1992) Association of agronomic traits with isozyme loci in perennial ryegrass (*Lolium perenne* L.). Euphytica 59: 141–150

Humphreys MW, Dalton SJ (1991) Stability at the phosphogluco-isomerase (PGI/2) locus in *Festuca arundinacea* plants regenerated from cell suspension and protoplast culture. Genome 34: 59–65

Humphreys MW, Dalton SJ (1992) Stability at the phosphoglucoisomerase (PGI/2) locus in *Lolium multiflorum* (2n = 4x = 28) × *Festuca arundinacea* (2n = 6x = 42) plants regenerated from cell suspension. Genome 35: 461–467

Kiang AS, Connolly V, McConnell DJ, Kavanagh TA (1993) Cytoplasmic male sterility (CMS) in *Lolium perenne* L.: 1. Development of a diagnostic probe for the male-sterile cytoplasm. Theor Appl Genet 86: 781–787

Lallemand J, Michaud O, Greneche M (1991) Electrophoretical description of ryegrass varieties: a catalogue. Plant Variet Seeds 4: 11–16

Liu CJ, Witcombe JR, Pittaway TS, Nash M, Hash CT, Busso CS, Gale MD (1994) An RFLP-based genetic map of pearl millet (*Pennisetum glaucum*). Theor Appl Genet 89: 481–487

Livesey V, Norrington-Davies J (1991) Isozyme polymorphism in *Festuca rubra* L. Euphytica 55: 73–79

Mazzucato A, Barcaccia G, Pezzotti M, Falcinelli M (1995) Biochemical and molecular markers for investigating the mode of reproduction in the facultative apomict *Poa pratensis* L. Sex Plant Reprod 8: 133–138

McMillin DE (1983) Plant isozymes: a historical perspective. In: Tanksley SD, Orton TJ (eds) Isozymes in plant genetics and breeding, part A. Elsevier, Amsterdam, pp 3–13

Moore GA, Collins GB (1983) New challenges confronting plant breeders. In: Tanksley SD, Orton TJ (eds) Isozymes in plant genetics and breeding, part A. Elsevier, Amsterdam, pp 25–58

Nielsen G (1980) Identification of all genotypes in tetraploid ryegrass (*Lolium* spp.) segregating for four alleles in a *Pgi*-enzyme locus. Hereditas 92: 49–52

Nielsen G (1985) The use of isozymes as probes to identify and label plant varieties and cultivars. In: Rarrazzi MC, Scandalios JG, Whitt GS (eds) Isozymes: current topics in biological and medical research. Liss, New York, pp 1–32

Nielsen G, Østergaard H, Johansen H (1985) Cultivar identification by means of isoenzymes II. Genetic variation at four enzyme loci in diploid ryegrass. Z Pflanzenzüchtg 94: 74–86

Østergaard H, Nielsen G (1981) Cultivar identification by means of isoenzymes. I-Genotypic survey of the *Pgi*-2 locus in tetraploid ryegrass. Z Pflanzenzüchtg 87: 121–132

Østergaard H, Nielsen G, Johansen H (1985) Genetic variation in cultivars of diploid ryegrass, *Lolium perenne* and *L. multiflorum*, at five enzyme systems. Theor Appl Genet 69: 409–421

Ozias-Akins P, Lubbers EL, Hanna WW, McNay JW (1993) Transmission of the apomictic mode of reproduction in *Pennisetum*: co-inheritance of the trait and molecular markers. Theor Appl Genet 85: 632–638

Pérez-Vicente R, Petris L, Osusky M, Potrykus I, Spangenberg G (1992) Molecular and cytogenetic characterization of repetitive DNA sequences from *Lolium* and *Festuca*: applications in the analysis of *Festulolium* hybrids. Theor Appl Genet 84: 145–154

Polans NO, Allard RW (1985) Inheritance of electrophoretically detectable variants in ryegrass. J Heredity 76: 61–62

Rafalski JA, Tingey SV (1993) Genetic diagnostics in plant breeding: RAPDs, microsatellites and machines. Trends Genet 9: 275–280

Rafalski A, Tingey S, Williams JGK (1994) Random amplified polymorphic DNA (RAPD) markers. In: Gelvin SB, Schilperoort RA (eds) Plant Mol Biol Manu. Kluwer Academic, Dordrecht, pp H4: 1–8

Rothe G (1994) Electrophoresis of enzymes: laboratory methods. Springer, Berlin Heidelberg New York, pp 1–15

Rouwendal GJA, Creemers-Molenaar J, Krens FA (1992) Molecular aspects of cytoplasmic male sterility in perennial ryegrass (*Lolium perenne* L.): mtDNA and RNA differences between plants with male-sterile and fertile cytoplasm and restriction mapping of their *atp6* and *coxI* homologous regions. Theor Appl Genet 83: 330–336

Shenoy VB, Vasil IK (1992) Biochemical and molecular analysis of plants derived from embryogenic tissue cultures of napier grass (*Pennisetum purpureum* K. Schum.). Theor Appl Genet 83: 947–955

Spangenberg G, Vallés MP, Wang ZY, Montavon P, Nagel J, Potrykus I (1994) Asymmetric somatic hybridization between tall fescue (*Festuca arundinacea* Schreb.) and irradiated Italian ryegrass (*Lolium multiflorum* Lam.) protoplasts. Theor Appl Genet 88: 509–519

Spangenberg G, Wang ZY, Legris G, Montavon P, Takamizo T, Pérez-Vicente R, Vallés MP, Nagel J, Potrykus I (1995) Intergeneric symmetric and asymmetric somatic hybridization in *Festuca* and *Lolium*. Euphytica 85: 235–245

Spoor W, Hay JM (1979) Identification of *Poa pratensis* lines using electrophoresis of seed extracts. Seed Sci Technol 7: 467–474

Stammers M, Harris J, Evans GM, Hayward MD, Forster JW (1995) Use of random PCR (RAPD) technology to analyse phylogenetic relationships in the *Lolium/Festuca* complex. Heredity 74: 19–27

Takamizo T, Spangenberg G, Suginobu K, Potrykus I (1991) Intergeneric somatic hybridization in Gramineae: somatic hybrid plants between tall fescue (*Festuca arundinacea* Schreb.) and Italian ryegrass (*Lolium multiflorum* Lam.). Mol Gen Genet 231: 1–6

Tanksley SD, Young ND, Paterson AH, Bonierbale MW (1989) RFLP mapping in plant breeding: new tools for an old science. Bio/Technology 7: 257–264

Thomas HM, Morgan WG, Meredith MR, Humphreys MW, Thomas H, Leggett JM (1994) Identification of parental and recombined chromosomes in hybrid derivatives of *Lolium multiflorum* × *Festuca pratensis* by genomic in situ hybridization. Theor Appl Genet 88: 909–913

Thomas HM, Harper JA, Meredith MR, Morgan WG, King IP (1997) Physical mapping of ribosomal DNA sites in *Festuca arundinacea* and related species by in situ hybridization. Genome 40: 406–410

Vallés MP, Wang ZY, Montavon P, Potrykus I, Spangenberg G (1993) Analysis of genetic stability of plants regenerated from suspension cultures and protoplasts of meadow fescue (*Festuca pratensis* Huds.). Plant Cell Rep 12: 101–106

Wang ZY, Nagel J, Potrykus I, Spangenberg G (1993b) Plants from cell suspension-derived protoplasts in *Lolium* species. Plant Sci 94:179–193

Waugh R, Powell W (1992) Using RAPD markers for crop improvement. TIBTECH 10: 186–191

Weaver KR, Callahan LM, Caetano-Anollés G, Gresshoff PM (1995) DNA amplification fingerprinting and hybridization analysis of centipedegrass. Crop Sci 35: 881–885

Weeden NF, Emmo AC (1985) Isozyme characterization of Kentucky bluegrass cultivars. Can J Plant Sci 65: 985–994

Wehner DJ, Duich JM, Watschke TL (1976) Separation of Kentucky bluegrass cultivars using peroxidase isoenzyme banding patterns. Crop Sci 16: 475–480

Welsh J, McClelland M (1990) Fingerprinting genomes using PCR with arbitrary primers. Nucleic Acid Res 18: 7213–7218

Williams JGK, Kubelik AR, Livak KJ, Rafalski JA, Tingey SV (1990) DNA polymorphisms amplified by arbitrary primers are useful as genetic markers. Nucleic Acid Res 18: 6531–6535

Wu L, Jampates R (1986) Chromosome number and isoenzyme variation in Kentucky bluegrass cultivars and plants regenerated from tissue culture. Cytologia 51: 125–132

Wu L, Harivandi AH, Harding JA, Davis WB (1984) Identification of Kentucky bluegrass cultivars with esterase and phosphoglucomutase isoenzyme markers. Crop Sci 24: 763–768

Xu WW, Sleper DA (1991) A survey of restriction fragment length polymorphisms in tall fescue and its relatives. Genome 34: 686–692

Xu WW, Sleper DA (1994) Phylogeny of tall fescue and related species using RFLPs. Theor Appl Genet 88: 685–690

Xu WW, Sleper DA, Chao S (1992) Detection of RFLP in perennial ryegrass, using heterologous probes from tall fescue. Crop Sci 32: 1366–1370

Xu WW, Sleper DA, Krause GF (1994) Genetic diversity of tall fescue germplasm based on RFLPs. Crop Sci 34: 246–252

Yamamoto I, Duich JM (1994) Electrophoretic identification of cross-pollinated bentgrass species and cultivars. Crop Sci 34: 792–798

Chapter 10
Perspectives

10.1 Introduction

Grassland agriculture is highly dependent upon a reliable source of forage as the primary feed base for ruminant livestock (Barnes and Baylor 1995). In most areas of the world, forage production is a low cash-input production system, and thus the most economical way to deliver advanced technology to forage producers is through the genetic improvement of forage plant cultivars (McKersie 1997). Conventional forage plant breeding has been based on the use of natural genetic variation as found between and within ecotypes or created through sexual recombination (Van Wijk et al. 1993). Biotechnology allows the generation of novel variability as well as more efficient use of existing genetic variability. It offers opportunities to enhance the sources of useful genes accessible for the development of new cultivars and to speed up genetic improvement programs.

In previous chapters, we have described the methodical basis for the genetic manipulation at the cellular level of key forage and turf grass species, mainly belonging to the *Festuca-Lolium* complex, in particular: (1) the establishment of plant regeneration systems from in vitro cultures (Chaps. 2–5); (2) the combination of whole or partial genomes by somatic hybridization (Chap. 6); (3) genetic transformation using selectable marker and reporter genes (Chaps. 7 and 8); and (4) the development of molecular markers (Chap. 9).

The challenge now is how to best deploy these molecular technologies and tools, to dissect the genetic basis of complex quantitative traits and follow up with marker-assisted selection, to evaluate the potential of experimental approaches based on the transgenic transfer of single valuable genes, to generate novel genetic variability and novel elite germplasm, and to efficiently incorporate these factors into breeding programs for the development of improved cultivars.

Primary target traits for molecular breeding of forage and turf grasses are forage quality, disease and pest resistance, enhanced persistency, tolerance to abiotic stresses, and the manipulation of growth and development. Some representative approaches are discussed below.

10.2 Forage Quality

Molecular breeding to overcome limitations in forage quality may be targeted to the individual subcharacters involved: dry matter digestibility, water-soluble carbohydrate content, and protein content. Molecular breeding approaches may thus include modification of the lignin profile to enhance dry matter digestibility, manipulation of fructan metabolism to increase non-structural carbohydrate content, and the expression of "rumen by-pass" proteins to improve the supply of proteins and essential amino acids. Since most quality or anti-quality parameters are associated with specific metabolic pathways or the production of specific proteins, target enzymes or suitable foreign proteins could be identified, corresponding genes isolated, and their expression manipulated in transgenic grass plants.

10.2.1 Manipulation of Lignin Biosynthesis

Dry matter digestibility of forage grasses declines markedly (>10%) as plants flower and senesce (Buckner et al. 1967; Radojevic et al. 1994; Stone 1994). The changes in dry matter digestibility greatly contribute to the lowering of nutritive value of grass forage during summer (Stone 1994). Increasing dry matter digestibility has been ranked as the most important goal in genetic improvement of nutritive value of forage grasses for dairy pastures (Smith et al. 1997). However, since heritability of dry matter digestibility is low and a large number of genes control it, the potential for rapid genetic improvement by traditional methods is low (Barnes 1990).

Lignification of plant cell walls has been identified as the major factor responsible for lowering digestibility of forage tissues as they mature (Buxton and Russell 1988). The inhibitory effects of lignin on forage digestibility depend on lignin monomer composition and functional groups, lignin content, and the extent of cross-linking to cell wall polysaccharides (Jung and Vogel 1986; Sewalt et al. 1997). Small increases in digestibility are expected to have significant impact on forage quality and concomitantly on animal productivity.

Lignification is comprised of a highly coordinated and regulated set of metabolic events resulting in the biosynthesis of lignin (Fig. 10.1; Whetten and Sederoff 1995; Boudet and Grima-Pettenati 1996; Boudet et al. 1996; Campbell and Sederoff 1996; Dixon et al. 1996).

Molecular breeding for improved digestibility by down-regulating lignin biosynthetic enzymes through antisense and sense suppression in transgenic plants is being explored. The main target enzymes being considered are caffeic acid *O*-methyltransferase (COMT), cinnamyl alcohol dehydrogenase (CAD), 4-coumarate:CoA ligase (4CL) and cinnamoyl CoA reductase (CCR). Experiments with model plant tobacco and poplar have shown that the down-regulation of COMT, CAD and 4CL expression leads to altered lignin

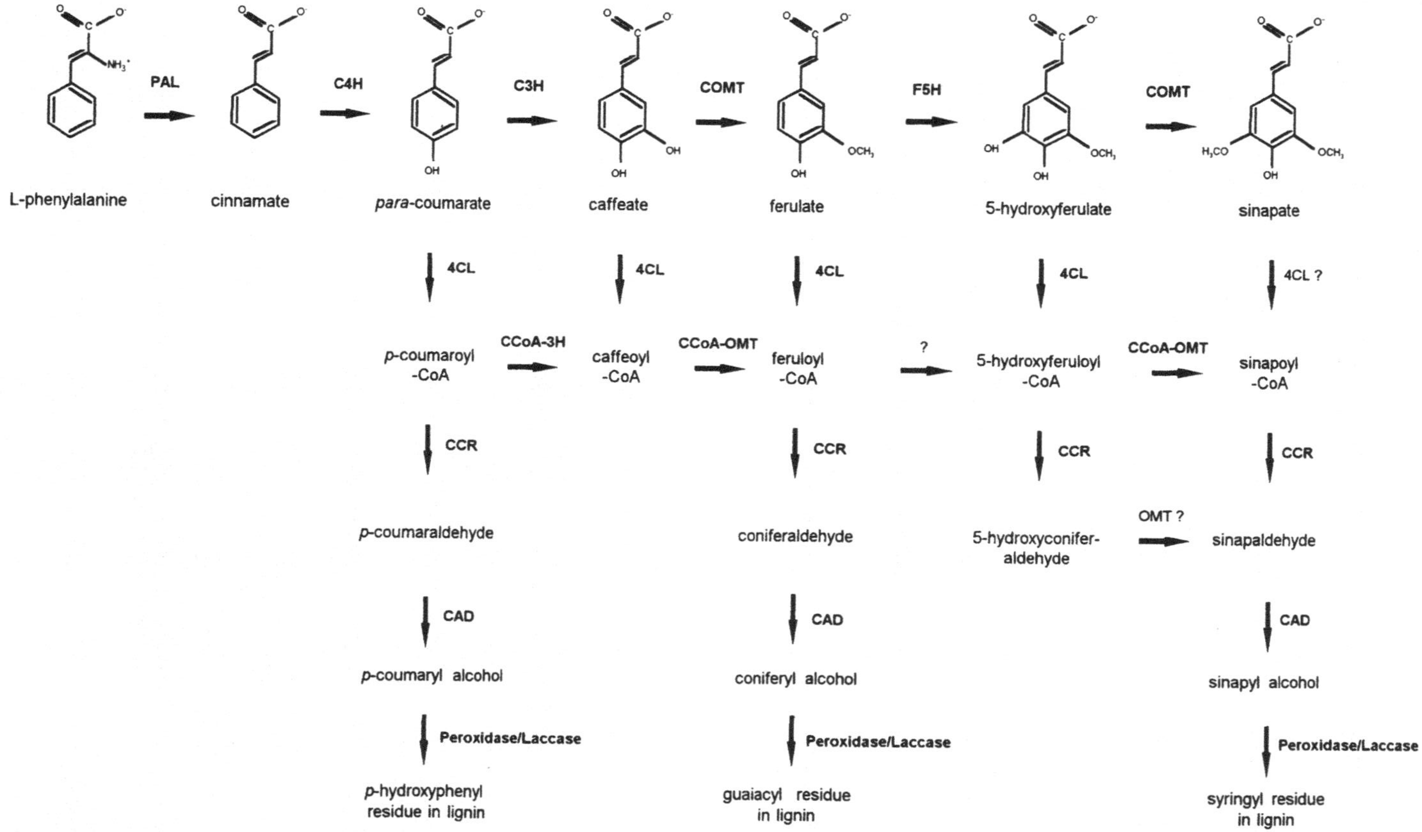

Fig. 10.1. The lignin biosynthetic pathway. The lignin biosynthetic enzymes are: *PAL*, phenylalanine ammonia-lyase; *C4H* cinnamate 4-hydroxylase; *C3H* 4-hydroxycinnamate 3-hydroxylase; *COMT* caffeic acid 3-*O*-methyltransferase; *F5H* ferulate 5-hydroxylase; *4CL* 4-coumarate: CoA ligase; *CCoA-3H* coumaroyl-coenzyme A 3-hydroxylase; *CCoA-OMT* caffeoyl-coenzyme A *O*-methyltransferase; *CCR* cinnamoyl-CoA reductase; *CAD* cinnamyl alcohol dehydrogenase. (Modified from Boudet et al. 1996; Campbell and Sederoff 1996)

composition or reduced lignin content (Table 10.1). A significant improvement in dry matter digestibility of transgenic sense and antisense COMT tobacco plants showing reduced COMT activity has been demonstrated (Bernard Vailhé et al. 1996a; Sewalt et al. 1997a). A decrease in syringyl lignin units (Bernard Vailhé et al. 1996a) or a reduction in lignin content (Sewalt et al. 1997a) was observed. Similarly, transgenic antisense CAD tobacco plants with down-regulated CAD activity have been reported to produce chemically more extractable lignin which was altered in composition (Halpin et al. 1994) or structure (Bernard Vailhé et al. 1996b). These results indicate that improvement of dry matter digestibility by the introduction of chimeric sense and antisense lignin biosynthetic genes can be achieved apparently without impairing normal development of the plant.

The technologies enabling the recovery of transgenic forage grasses have opened up opportunities for their molecular breeding for improved digestibility and forage quality. Perennial ryegrass cDNA homologues of COMT (Fig. 10.2) have been isolated and characterized (Heath et al. 1997). Corresponding chimeric genes are currently being used to down-regulate the expression of COMT in transgenic ryegrass and fescue plants. Once proof of concept for the transgenic forage grasses is obtained, thorough agronomic assessment of transgenic plants and hybridization to generate transgenic elite germplasm with subsequent selection will be required to produce marketable cultivars.

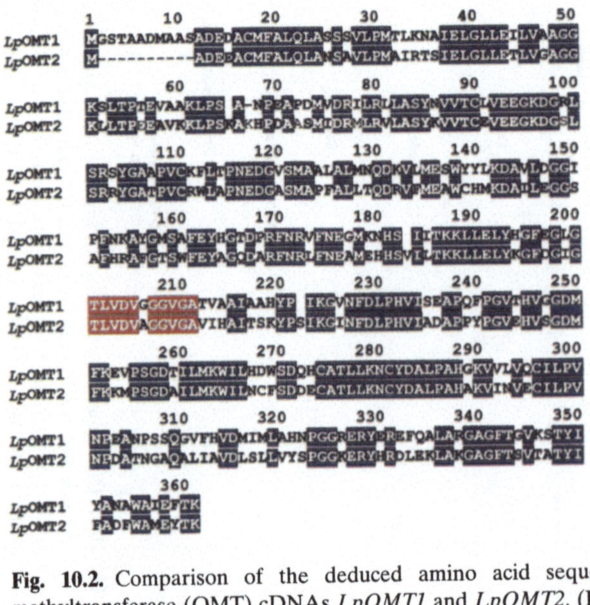

Fig. 10.2. Comparison of the deduced amino acid sequences of perennial ryegrass *O*-methyltransferase (OMT) cDNAs *LpOMT1* and *LpOMT2*. (Heath et al. 1997)

Table 10.1. Genetic manipulation of lignin biosynthesis in plants

Target enzyme	Transgene origin	Transgenic plant	Approach	Lignin content	Lignin composition/ structure	S/G ratio	Degradability/ extractability	Source
PAL	Bean	Tobacco	Sense	Reduced	Altered, methoxyl↑	–	Increased	Bate et al. (1994)
C4H	Alfalfa	Tobacco	Sense/ antisense	Reduced	Altered, S↓	Decreased	–	Sewalt et al. (1997a) Sewalt et al. (1997b)
COMT	Alfalfa	Tobacco	Antisense	Reduced	Unaltered	Unchanged	–	Ni et al. (1994)
COMT	Poplar	Tobacco	Antisense	Unchanged	Altered, S↓	Decreased	–	Dwivedi et al. (1994)
COMT	Poplar	Poplar	Antisense	Unchanged	Altered, S↓ G↑	Decreased	–	Van Doorsselaere et al. (1995)
COMT	Tobacco	Tobacco	Sense/ antisense	Unchanged	Altered, S↓	Decreased	–	Atanassova et al. (1995)
COMT	Tobacco	Tobacco	Sense/ antisense	Unchanged	Altered, S↓	Decreased	Increased	Bernard Vailhé et al. (1996a)
COMT	Alfalfa	Tobacco	Antisense	Reduced	Altered, S↓ G↓	Increased/ unchanged	Increased	Sewalt et al. (1997a)
4CL	Tobacco	Tobacco	Sense/ antisense	Reduced	Altered, S↓ G↓	Varied	–	Kajita et al. (1996)
CAD	Tobacco	Tobacco	Antisense	Unchanged	Altered, cinnamaldehyde↑	Decreased	Increased	Halpin et al. (1994)
CAD	*Aralia cordata*	Tobacco	Antisense	Unchanged	Altered, cinnamaldehyde↑	Increased	–	Higuchi et al. (1994) Hibino et al. (1995)
CAD	Poplar	Poplar	Sense/ antisense	Unchanged	Altered, aldehydes↑	Unchanged	Increased	Baucher et al. (1996)
CAD	Tobacco	Tobacco	Antisense	Unchanged	Altered structure	Unchanged	Increased	Bernard Vailhé et al. (1996b)
CAD	Tobacco	Tobacco	Antisense	Unchanged/ increased	Altered, cinnamaldehyde↑	–	Increased	Stewart et al. (1997)
PER	Tobacco	Tobacco	Ribozyme/ antisense	Unchanged	–	–	–	McIntyre et al. (1996)
PER	Tobacco	tobacco	Antisense	Unchanged	–	–	–	Lagrimini et al. (1997)
PER	Tobacco	Tobacco	Ribozyme/ antisense	Unchanged	–	–	–	McIntyre et al. (1996)

PAL, phenylalanine ammonia-lyase; COMT, caffeic acid *O*-methyltransferase; 4CL, 4-coumarate: CoA ligase; CAD, cinnamyl alcohol dehydrogenase; PER, peroxidase; G, guaiacyl lignin unit; S, syringyl lignin unit; ↑, increased; ↓, decreased

10.2.2 Manipulation of Fructan Metabolism

Fructans are polyfructose molecules produced by many grass species as their main soluble storage carbohydrate form. It has been shown that ryegrass lines which accumulate higher concentrations of soluble carbohydrates do not suffer as great a decline in digestibility during summer (Radojevic et al. 1994). The increased level of soluble carbohydrates appears to offset the decline in digestibility due to lignification. In addition, herbage intake, protein capture in the rumen and live-weight gains may be improved by increasing the concentrations of non-structural carbohydrates in pasture plants (Michell 1973; Jones and Roberts 1991; Beever 1993).

The introduction of the microbial fructosyltransferase *sacB* gene from *Bacillus subtilis* into fructan-devoid and starch-accumulating tobacco and potato plants led to the accumulation of considerable amounts of high molecular weight levan-type fructans (Ebskamp et al. 1994; Van der Meer et al. 1994). These results demonstrate that sucrose, the substrate for fructosyltranferase, can be efficiently routed into a new sink in non-fructan-accumulating species. Furthermore, transgenic tobacco plants that accumulate bacterial levan showed enhanced performance under drought stress (Pilon-Smits et al. 1995). The sucrose:fructan 6-fructosyltransferase (6-SFT) from barley, a fructan-accumulating species, has been purified and the corresponding cDNA isolated (Sprenger et al. 1995). Transgenic ryegrass plants with altered fructan metabolism brought about by the expression of chimeric bacterial levansucrase genes have been generated (Ye 1997). The isolation and characterization of grass fructan biosynthetic genes and the manipulation of their expression in transgenic plants will enhance our knowledge of fructan metabolism and carbohydrate partitioning in grasses and clarify their functional role in tolerance to cold and drought. This knowledge will be instrumental in designing experimental approaches to producing transgenic lines with enhanced forage quality and tolerance to abiotic stresses.

10.2.3 Transgenic Expression of "Rumen By-pass" Proteins

Wool growth in sheep is limited by the supply of sulphur-containing amino acids such as methionine and cysteine (Reis and Schinckel 1963). Direct infusion into the abomasum by-passing the rumen of cysteine alone or its precursor methionine may lead to an increase in wool growth of up to 100% (Reis 1988). It is expected that the ingestion of forage containing relatively rumen-stable proteins rich in sulphur-containing amino acids may lead to an enhanced supply of limiting essential amino acids for sheep nutrition and thus to increased wool growth (Higgins et al. 1989; Rogers 1990).

The production of transgenic forage legumes expressing genes encoding different "rumen by-pass" proteins rich in sulphur-containing amino acids, e.g. chicken ovalbumin, pea albumin and sunflower seed albumin, has been reported (Schroeder et al. 1991; Ealing et al. 1994; Tabe et al. 1995; Khan et al. 1996). These "rumen by-pass" proteins accumulated at levels from <0.01 to 0.3% of total extractable protein (Schroeder et al. 1991; Ealing et al. 1994; Tabe et al. 1995; Khan et al. 1996). Transgenic forage grass plants expressing sunflower seed albumin have also been obtained (Wang et al., in preparation). Chimeric genes carrying sunflower albumin SFA8 coding sequence (Kortt et al. 1991) including the endoplasmic reticulum retention signal KDEL (Wandelt et al. 1992) under control of different promoters were generated and corresponding transgenic tall fescue plants recovered through biolistic

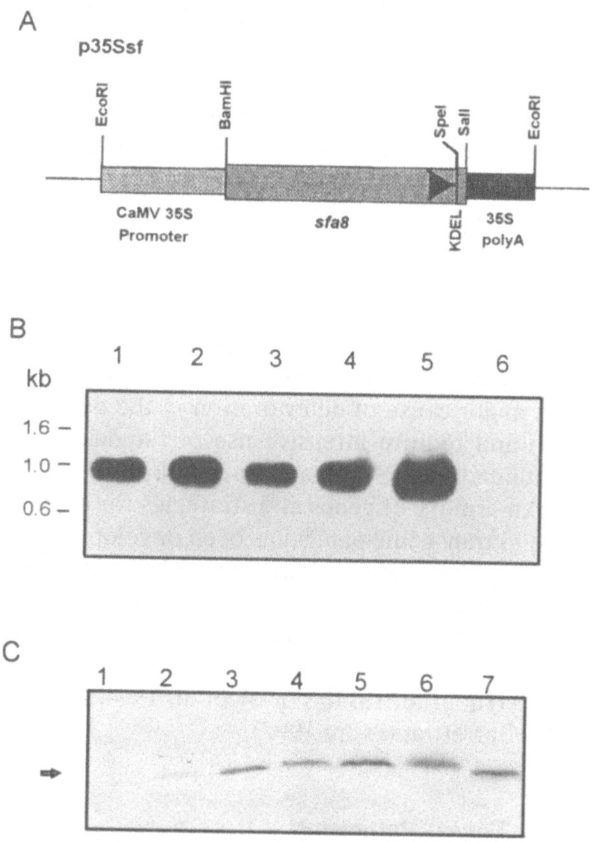

Fig. 10.3A–C. Expression of a chimeric sunflower albumin gene in transgenic tall fescue (*Festuca arundinacea*) plants. **A** Map of plasmid p35Ssf bearing a chimeric sunflower albumin gene (*sfa8*) including the *KDEL* endoplasmic reticulum-retention signal used for genetic transformation of tall fescue. **B** Northern hybridization analysis of *sfa8*-transgenic tall fescue plants. The *sfa8* sequence was used as hybridization probe. *Lanes 1–5* Five independent *sfa8*-transgenic tall fescue plants; *6* untransformed tall fescue plant as negative control. **C** Western blot analysis of *sfa8*-transgenic tall fescue plants. A goat anti-sunflower albumin antibody was used. 80 μg of soluble leaf protein were loaded in each lane. *Lane 1* Untransformed tall fescue plant as negative control; *2–6* five independent *sfa8*-transgenic tall fescue plants; *7* 80 μg of SFA8 protein

transformation (Wang et al., in preparation). The transgenic tall fescue plants produced the expected *sfa8* transcript, and the corresponding methionine-rich protein accumulated at levels of up to 0.2% of total soluble protein (Fig. 10.3). In order to achieve nutritionally useful levels, expression of the sunflower seed albumin may be required to reach 2–5% of total soluble protein. Strategies for increasing the accumulation levels of foreign proteins in leaves of forage plants may include the use of stronger promoters and the development of transplastomic lines. Transplastomic lines with transformed chloroplasts have been shown to accumulate such expression levels of foreign protein as mentioned above. Amplification of a *Bacillus thuringiensis* (Bt) gene in chloroplasts resulted in the accumulation of an unprecedented 3–5% of the soluble protein in transplastomic tobacco leaves (McBride et al. 1995). Chloroplast transformation for and the recovery of transplastomic grasses and cereals still remains to be demonstrated. Chloroplast transformation and transplastomic plants may prove suitable systems for transgene containment in outcrossing grass species showing maternal inheritance of chloroplasts.

10.3 Disease and Pest Resistance

Pathogen and pest infection can considerably lower herbage yield, persistency, nutritive value, and palatability of forage grasses (Reed 1994). Fungal diseases are a major cause of deterioration in the quality of amenity turf (Smith et al. 1989) and require intensive use of fungicides in the management of highly maintained turf (Schumann and Wilkinson 1992).

An armory of genes and strategies for engineering disease and pest resistance in transgenic plants has been developed and tested over the last decade, including chitinases, glucanases, plant defensins, phytoalexins, ribosome-inactivating proteins, viral coat proteins, Bt toxins, proteinase inhibitors, and α-amylase inhibitors. Some of them have been applied to the development of pasture plants, mainly forage legumes, for enhanced disease and pest resistance (Hill et al. 1991; Voisey et al. 1994; Masoud et al. 1996; Strizhov et al. 1996; Garrett and Chu 1997).

10.3.1 Fungal Pathogens

Fungi attack leaves and root systems of turf grass causing leaf and root damage which results in unsightly greens and poor establishment and persistency. The fungal pathogens *Rhizoctonia solani* and *Drechslera dictyoides* are major problems in turf where they cause brown patch and net blotch, respectively. A number of *Pythium* species can also cause devastation to seedlings and mature turf under hot and humid conditions. Cold, moist conditions favour the development of winter Fusarium or pink snow mould, *Microdochium nivale*, which causes the disease complex Fusarium patch. Conventional breeding approaches of introducing resistance to the main fungal diseases into turf grasses are difficult since very few resistance sources are available (Meyer 1982; Meyer and Funk 1989).

The constitutive, organ-specific or pathogen-inducible expression in transgenic plants of genes encoding antifungal proteins and acting either individually or in a concerted manner may confer useful novel types of resistance to fungal diseases. Transgenic tobacco and canola plants expressing a bean endochitinase (*chi5B*) gene have shown increased protection against *R. solani*. (Broglie et al. 1991; Benhamou et al. 1993). The chimeric *chi5B* gene has also been introduced into fine-leaved turf fescues by biolistic transformation, and increased levels of chitinase activity were observed in the transgenic plants recovered (Fig. 10.4). Transgenic approaches pyramiding different resistant genes may give substantially greater protection against fungal pathogens than the deployment of individual synthetic resistance genes (Zhu et al. 1994).

The soil-borne fungus *Microdochium nivale* causes snow mould in ryegrasses and leads to winter killing in regions with prolonged snow cover (Arsvoll 1973; Matsumoto and Araki 1982; Winter 1986). Forage grass plants with the superior nutritive value of the ryegrasses, yet capable of persisting under less favourable conditions are needed. Red fescue (*Festuca rubra*) is a rhizomatous grass which can persist under such conditions and is much less susceptible to snow mould but has lower quality than the ryegrasses. Intergeneric asymmetric somatic hybrids have been obtained by fusion of metabolically inactivated protoplasts of *F. rubra* and X-ray-irradiated protoplasts of *L. perenne* (Fig. 10.5), and their genomic make-up has been characterized using species-specific repetitive DNA sequences (Fig. 10.6; Legris 1996). This asymmetric somatic hybridization approach might create hybrids useful as bridges for further breeding and to overcome difficulties encountered in intergeneric crosses between *L. perenne* and *F. rubra* (Jenkin 1933, 1955).

10.3.2 Viruses

Barley yellow dwarf virus (BYDV) and ryegrass mosaic virus (RMV) are widely distributed in ryegrasses and fescues (Chapman 1979; Catherall and Parry 1987; Henry and Dedryver 1991; Eagling et al. 1992). BYDV infection in ryegrasses reduces dry matter yield by as much as 24% (Catherall 1966; Wilkins and Catherall 1977). RMV infection in ryegrasses results in dry matter reductions of 5–50% (A'Brook and Heard 1975; Jones et al. 1977; Eagling et al. 1992). RMV infection also reduces the competitiveness of perennial ryegrass, resulting in poor establishment and reduced persistence (Eagling et al. 1991).

Coat protein-mediated transgenic resistance has proven to be a valuable means of protecting plants from certain virus infections. This most widely reported transgenic approach against virus infection can be applied to the development of transgenic ryegrasses with enhanced resistance to ryegrass mosaic and barley yellow dwarf viruses.

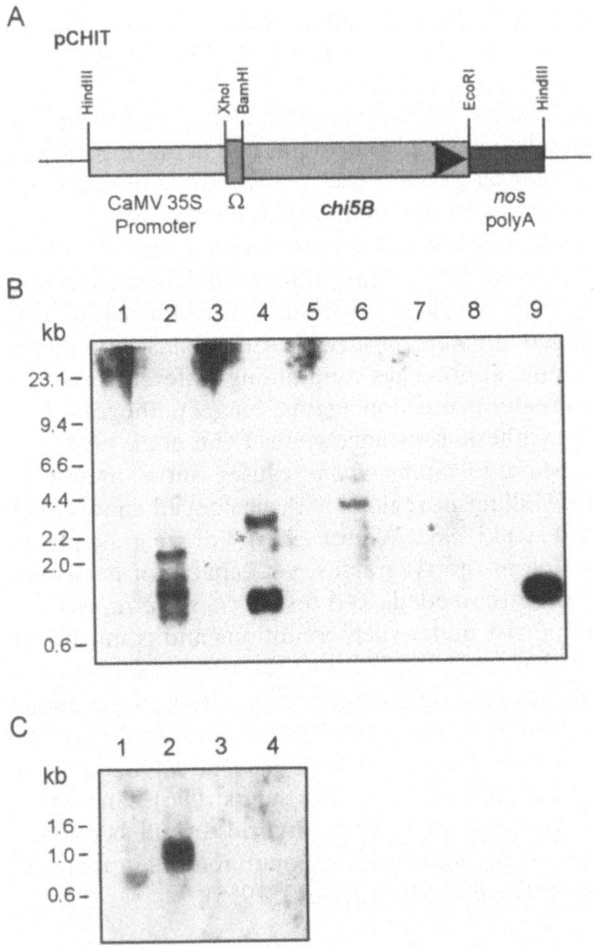

Fig. 10.4A–C. Expression of a chimeric endochitinase gene in transgenic fine-leaved turf red fescue (*Festuca rubra*) plants. **A** Map of plasmid pCHIT bearing a chimeric bean endochitinase gene (*chi5B*) used for genetic transformation of turf red fescue. **B** Southern blot with undigested (*odd-numbered lanes*) and *Bam*HI- and *Eco*RI-digested (*even-numbered lanes*) total cellular DNA samples isolated from transgenic red fescue plants hybridized with a *chi5B* probe. *Lanes 1–6* Three independent transgenic red fescue plants; *7–8* untransformed red fescue plant as negative control; *9* 10 pg of *Bam*HI- and *Eco*RI-digested pCHIT as positive control. **C** Northern hybridization analysis of *chi5B*-transgenic red fescue plants. The *chi5B* cDNA was used as hybridization probe. *Lanes 1–3* Three independent *chi5B*-transgenic red fescue plants; *4* untransformed red fescue plant as negative control

Fig. 10.5A–I. Asymmetric somatic hybrid plants obtained from protoplast fusion between red fescue (*Festuca rubra*) and perennial ryegrass (*Lolium perenne*). **A** Embryogenic cell suspension of *F. rubra* used for isolation of totipotent protoplasts. **B** Non-morphogenic cell suspension of *L. perenne* used for isolation of protoplasts. **C** Protoplast mix after fusion of iodoacetamide-inactivated protoplasts of *F. rubra* and non-morphogenic protoplasts of *L. perenne* isolated from **A** and **B**, respectively. **D** Putative somatic hybrid colonies obtained from fusion experiments between red fescue protoplasts and 400-Gy-irradiated perennial ryegrass protoplasts after 4 weeks in bead-type culture. **E** Regenerating clone from **D** 2 months after protoplast fusion. **F** Rooted plantlets. **G** Greenhouse-growing mature somatic hybrid plants obtained from asymmetric protoplast fusions between *F. rubra* and *L. perenne* protoplasts. **H** Asymmetric somatic hybrid plants growing under field conditions. **I** Flowering asymmetric somatic hybrids growing under field conditions. (Legris 1996)

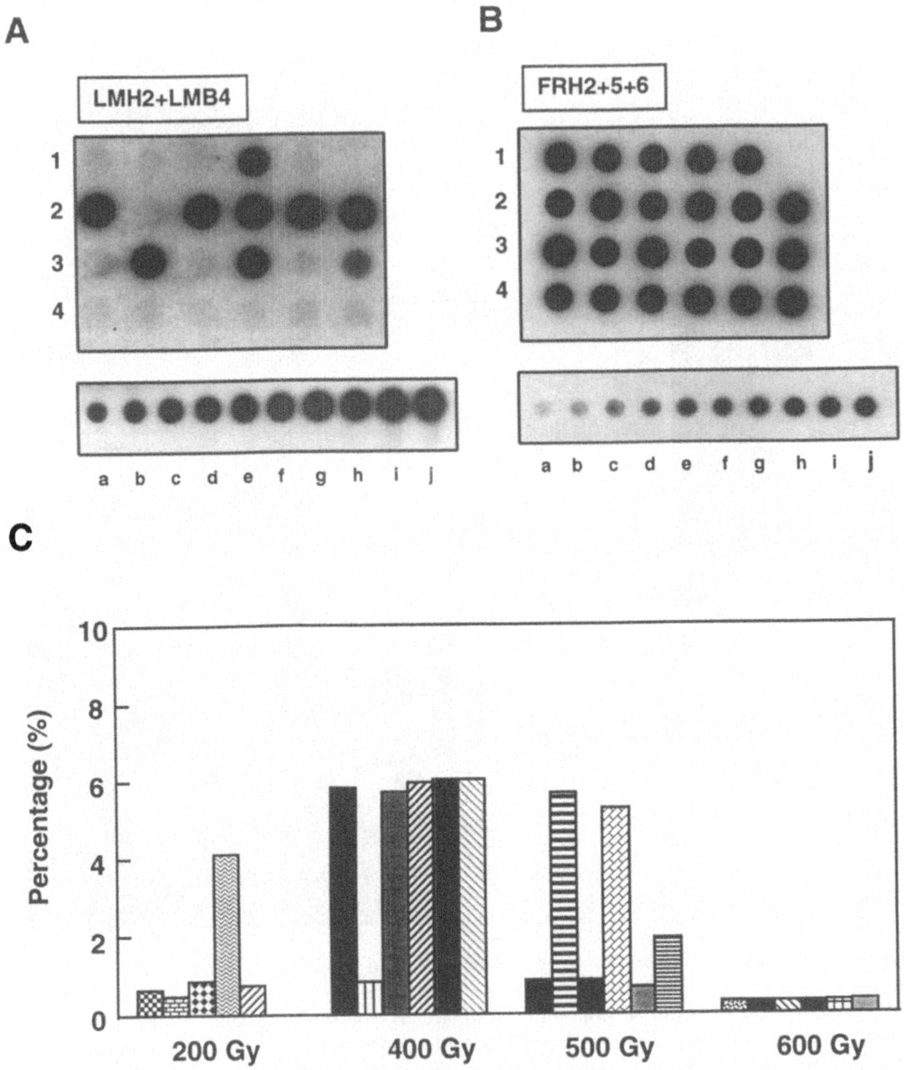

Fig. 10.6A–C. Nuclear composition of asymmetric *F. rubra* (+) *L. perenne* somatic hybrids. **A** Dot blot analysis using *L. multiflorum*-specific probes *LMB4* and *LMH2* of asymmetric somatic hybrid clones (calli and plants) obtained from independent fusion products between recipient protoplasts of *F. rubra* and donor protoplasts of *L. perenne* irradiated with *1* 200 Gy; *2* 400 Gy; *3* 500 Gy; and *4* 600 Gy. *a–j* Increasing amounts of *L. perenne* DNA: *a* 1 μg; *b* 2 μg; *c* 3 μg; *d* 4 μg; *e* 5 μg; *f* 6 μg; *g* 7 μg; *h* 8 μg; *i* 9 μg and *j* 10 μg for calibration. **B** Dot blot analysis using *F. rubra*-specific probes *FRH2* and *FRH5* and *FRH6* of asymmetric somatic hybrid clones obtained from independent fusion products between recipient protoplasts of *F. rubra* and donor protoplasts of *L. perenne* irradiated with *1* 200 Gy; *2* 400 Gy; *3* 500 Gy; and *4* 600 Gy. *a–j* Increasing amounts of *F. rubra* DNA: *a* 1 μg; *b* 2 μg; *c* 3 μg; *d* 4 μg; *e* 5 μg; *f* 6 μg; *g* 7 μg; *h* 8 μg; *i* 9 μg and *j* 10 μg for calibration. **C** Genome composition of asymmetric *F. rubra* (+) *L. perenne* somatic hybrids. Quantitative dot blots of the asymmetric somatic clones (*columns* correspond to independent regenerating calli and plants) obtained at different doses (200–600 Gy) of X-ray irradiation of donor protoplasts were hybridized with the specific sequences. Genomic composition of independent clones is shown as a percentage of *L. perenne* DNA per 100% *F. rubra* DNA. (Legris 1996)

10.3.3 Pests

Pasture pests can damage plants directly by consuming foliage and roots or indirectly by transmitting plant pathogens (Allen 1987). Expression of Bt toxins was shown to be effective against selected insect pests in transgenic forage legumes (Voisey et al. 1994; Strizhov et al. 1996). This approach can be adopted for protection of forage and turf grasses against target pests which can be controlled by Bt toxins. Surrogate transformation developed in grasses (Chap. 7) opened up opportunities for using "ruminant-safe" strains of the endophyte to express and secrete protective proteins such as Bt toxins and proteinase inhibitors in order to protect forage grasses from pest damage.

10.4 Growth and Development

10.4.1 Manipulation of Pollen Allergens

Hayfever and seasonal allergic asthma due to grass pollen are environmental diseases that afflict up to 25% of the population in cool temperate climates (Ong et al. 1993; Tamborini et al. 1995). Ryegrass pollen shows the greatest abundance of all types of grass pollen in cool temperate climates (Marsh 1975; Smart et al. 1979) and is the major allergen for 49–67% of allergic patients (Ford and Baldo 1986; Freidhoff et al. 1986; Knox et al. 1989).

The pollen of *L. perenne* contains at least four major classes of allergenic proteins, each composed of multiple, immunologically indistinguishable isoforms involving 17 allergens ranging in size from 12 to 89 kDa (Ford and Baldo 1986; Singh et al. 1991; Sidoli et al. 1993; Knox and Suphioglu 1996). At least one protein from each of these classes has been isolated and characterized in some detail. cDNA clones for the major ryegrass pollen allergens Lol p 1 (Perez et al. 1990), Lol p 2 (Sidoli et al. 1993) and Lol p 5 (Singh et al. 1991) have been isolated. A number of Lol p 1- and Lol p 2-related sequences exist in the genomes of both *L. perenne* and *L. multiflorum* (Fig. 10.7). Transgenic *L. perenne* and *L. multiflorum* plants bearing antisense Lol p 1 and Lol p 2 cDNAs under the control of a pollen-specific promoter have been generated (Wu et al. 1997). These plants will allow to study the functional role in planta of the main pollen allergens and to explore the potential for the generation of hypo-allergenic ryegrass cultivars (Spangenberg et al. 1997).

10.4.2 Manipulation of Flowering Time and Senescence

The decline of feeding value in perennial grasses is associated with the onset of stem growth, flowering and leaf senescence. Stopping the formation of the

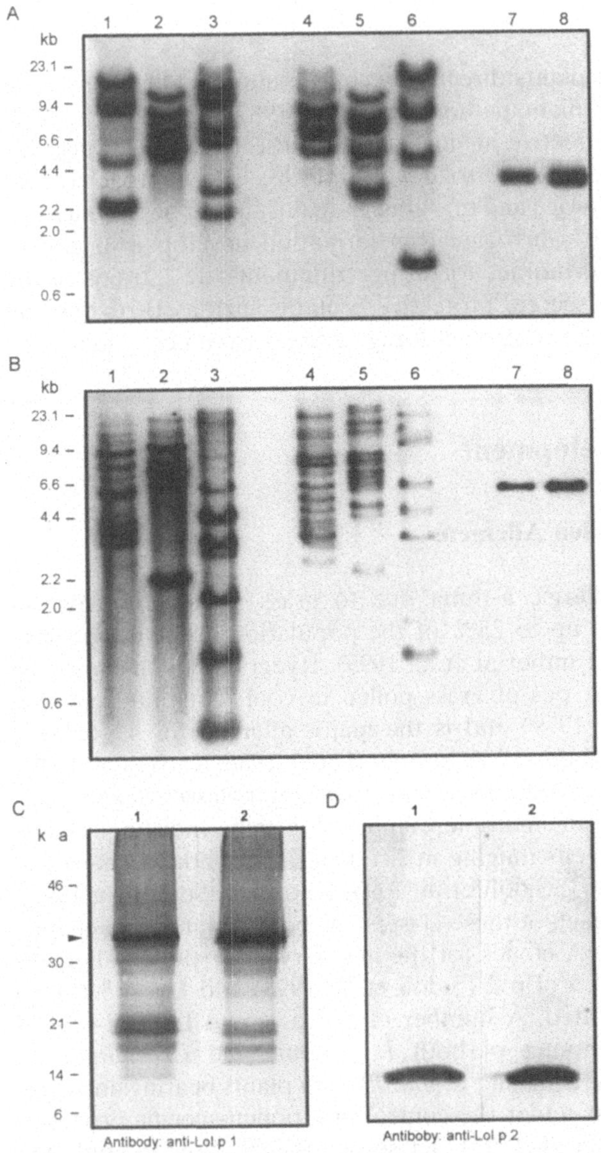

Fig. 10.7A–D. Pollen allergens in *Lolium perenne* and *L. multiflorum*. **A**,**B** Southern hybridization analyses for the detection of Lol p 1 (**A**) and Lol p 2 (**B**) ryegrass pollen allergen homologues in *L. perenne* and *L. multiflorum*. *Lanes 1–3 L. perenne* DNA digested by *Bam*HI, *Hin*dIII and *Eco*RI, respectively; *4–6 L. multiflorum* DNA digested by *Bam*HI, *Hin*dIII and *Eco*RI, respectively; *7–8* 5 and 10 pg of *Hin*dIII-digested plasmid pLol p 1 DNA (**A**) or pLol p 2 DNA (**B**). **C**, **D** Western blot analyses for the detection of Lol p 1 (**C**) and Lol p 2 (**D**) ryegrass pollen allergen homologues in *L. perenne* and *L. multiflorum* pollen protein extracts. Antibodies produced against recombinant perennial ryegrass Lol p 1 and Lol p 2 pollen allergens were used. *Lane 1 L. perenne* pollen protein extract; *2 L. multiflorum* pollen protein extract

less digestible stems or delaying the senescence process of leaves is expected to increase herbage quality.

Large modifications of flowering time in transgenic plants caused by regulating the expression of floral meristem initiation genes have been reported. The constitutive expression of the flower-meristem-identity genes *Leafy* or *Apetala1* of *Arabidopsis thaliana* has led to precocious flower development in transgenic aspen (Weigel and Nilsson 1995) and *A. thaliana* (Mandel and Yanofsky 1995; Weigel and Nilsson 1995). *A. thaliana* plants carrying mutations in one or more flower-meristem-identity genes exhibit either incomplete transformation from flowers to inflorescences or abolished flower development (Bowman et al. 1993). Thus, opportunities exist to control or inhibit flowering in transgenic perennial grasses by down-regulating the expression of genes determining floral meristem initiation, such as *Leafy* or *Apetala1*. Controlled or inhibited flowering in transgenic ryegrasses and fescues through sense or antisense suppression of *Leafy* and *Apetala1* homologues could lead to increased quality, improved seasonal growth patterns and transgene containment. For seed production, the flowering block of the non-flowering lines needs to be reversed. An inducible promoter for controlling the suppression of *Leafy* or *Apetala1* in the non-flowering transgenic grass plants, such as the copper-controllable gene expression system (Mett et al. 1993, 1996), may be considered.

Inhibition of leaf senescence has been achieved by the autoregulated production of cytokinin in transgenic plants (Gan and Amasino 1995). The system is based on the use of the *A. thaliana* senescence-specific SAG12 promoter controlling the transgenic expression of the isopentenyl transferase gene (*ipt*) of *Agrobacterium tumefaciens*, which catalyzes the rate-limiting step in cytokinin biosynthesis. Delayed leaf senescence but no developmental abnormalities were observed in transgenic tobacco plants that expressed the chimeric SAG12-*ipt* gene (Gan and Amasino 1995). Analogous chimeric *ipt* genes under control of grass leaf senescence-specific promoters may need to be tested to assess the suitability of this approach for developing forage grasses with enhanced leaf longevity.

10.4.3 Manipulation of Apomixis

Apomixis is the predominant mode of reproduction in several important forage grasses, where embryo formation solely from cells in the maternal ovule tissues rather than from the fusion of male and female gametes takes place (Hanna and Bashaw 1987). Since meiosis and fertilization are not involved in embryo formation, fertile seeds resulting from apomictic reproduction bear embryos genetically identical to the female parent plant (Bashaw and Hanna 1990; Koltunow 1993).

Genetic improvement of apomictic grasses is extremely difficult due to the inability to control variation in the progeny of facultative apomicts and the barriers to hybridization present in obligate apomicts (Akashi et al. 1993). In breeding apomictic forage grasses, there is an interest in overcoming the apomictic barrier to hybridization in order to generate new genotypes that can then be perpetuated, preserving even such characters as heterosis through successive generations via seeds (Akashi and Adachi 1992; Koltunow 1993). Somaclonal variation might contribute to the generation of genetic variability in apomictic grass species. Furthermore, intraspecific somatic hybridization by protoplast fusion may prove useful for creating novel variability in apomictic grasses. If the apomictic trait could be introduced into sexually reproducing crops, it would provide a source of renewable seed of high-yield potential, i.e. an agricultural trait of great value (Hanna and Bashaw 1987; Koltunow 1993). With the exception of pearl millet (Dujardin and Hanna 1989), programs aimed at transferring apomixis to agricultural crops on the basis of conventional breeding methods have failed. Interspecific or intergeneric asymmetric somatic hybridization may assist in transferring apomixis to crops of economic importance.

Studies on the apomictic mode of reproduction and the use of molecular markers are available for tropical grasses. A maize linkage group related to apomixis has been identified in *Brachiaria* (Pessino et al. 1997). Two cDNA fragments putatively related to apomixis were identified in *Brachiaria* on the basis of PCR differential display and non-radioactive detection using mRNA from pooled samples of apomictic and sexual plants (Leblanc et al. 1997). The isolation of genes responsible for the distinction between apomictic and sexual individuals will represent a significant step towards the exploitation of apomixis.

10.4.4 Manipulation of Self-Incompatibility and Cytoplasmic Male Sterility

Although *L. perenne* and *L. multiflorum* are normally self-incompatible (Jenkin 1931), self-compatible genotypes may occur relatively frequently (Forster and Wright 1970). Spontaneous mutations to self-compatibility have been identified in perennial ryegrass (Thorogood and Hayward 1991) and may be transferred to self-incompatible genotypes by recurrent selection. Furthermore, the self-fertile allele (Z_f) from the natural inbreeder *L. temulentum* may also be transferred to *L. perenne* and *L. multiflorum* through interspecific hybridization (Thorogood and Hayward 1988, 1992). The isolation of a putative S gene from *Phalaris coerulescens* (Li et al. 1994) has opened up opportunities for the cloning of corresponding *L. perenne* homologues and their down-regulation in transgenic ryegrasses.

The inter- and intraspecific transfer of the cytoplasmic male-sterile (CMS) character by somatic hybridization might help to establish a system for F_1 hybrid seed production in valuable genetic backgrounds. The transfer of the CMS trait to self-compatible inbred lines, particularly in ryegrasses, could contribute to further exploitation of the heterosis potential. Asymmetric somatic hybridization offers opportunities to generate the desired nucleus-cytoplasm combination in one step (Creemers-Molenaar et al. 1992). Similarly, intraspecific symmetric somatic hybridization in ryegrasses by protoplast fusion between different diploid genotypes could lead to tetraploid materials exploiting heterosis to a greater extent than present alternatives based on spontaneous or colchicine-induced chromosome doubling of individual diploid materials.

Different approaches on the manipulation of reproductive development – as outlined in Sections 10.4.2 to 10.4.4. – which may lead to a block of flowering, to the development of apomictic and male sterile transgenic wind-pollinated outcrossing grasses would provide for transgene contain-ment. Pollen dispersal is an important factor in the risk assessment of the release of genetically modified wind-pollinated grasses (Giddings et al. 1997a, b).

10.4.5 Grasses as Bioreactors

Plants can be used to express recombinant heterologous proteins. Transgenic plants may be an attractive alternative to microbial systems for the production of certain biomolecules (Goddijn and Pen 1995). The perennial growth habit, the biomass production potential, and the ability to grow in marginal areas exhibited by many grass species make them potential suitable candidates for molecular farming. Advances in genetic manipulation technologies that allow high levels of transgene expression and transgene containment may in the future make it possible to exploit some grass species as bioreactors for the production, among others, of industrial enzymes, pharmaceuticals, vaccines, antibodies and biodegradable plastics. Multidisciplinary efforts will, however, be needed to identify the most feasible targets, to generate transgenic plants with suitable expression levels, and to develop efficient downstream processing technology that could make transgenic grasses a cost-effective alternative for molecular farming.

10.5 Summary and Conclusions

Significant progress has been made in establishing the methodical basis for the genetic manipulation of the key grass species which form the foundation of grassland agriculture in temperate climates throughout the world. Reproducible and efficient plant regeneration systems that allow the recovery of largely fertile and genetically stable plants have been established. Symmetric and asymmetric protoplast fusion has led to the recovery of the first intergeneric somatic hybrid grass plants. Transgenic forage and turf grasses have been obtained by direct gene transfer to protoplasts, biolistic-mediated and whiskers-mediated transformation, and surrogate transformation with genetically modified endophytes. These methodologies have opened up opportunities to evaluate novel approaches to forage and turf grass improvement. A number of plant biotechnological approaches currently being tested include nutritional improvements through altered biosynthesis of lignin and soluble carbohydrates, the regulated expression of "rumen by-pass" proteins rich in essential amino acids, protection against pathogens and pests through engineered virus resistance and regulated expression of antifungal and pesticidal proteins, as well as the manipulation of growth and development aimed at improved persistency, improved tolerance to abiotic stresses, delayed senescence, non-flowering genotypes and down-regulation of pollen allergens.

Gaps in our understanding of the underlying genetics, physiology and biochemistry of many complex plant processes are likely to delay our progress in applying gene technology-based approaches to forage and turf grass improvement. However, the methodical basis currently available represents a very powerful tool to genetically manipulate these processes in order to enhance our understanding.

Molecular markers offer prospects of more precisely identifying superior individuals, enhancing selection efficiency, and as molecular tools in cultivar identification and marker-assisted selection. The use of markers is likely to have its greatest impact on the manipulation of quantitative characters for grass improvement. Future efforts in the development and deployment of molecular marker technology in grasses are thus expected to be focussed on the identification and location of quantitative trait loci by the presence of one or more linked markers.

The application of molecular methodologies and tools in forage and turf grass improvement will enhance current empirical phenotype-based selection with more directed and predictable genotype-based approaches. However, these molecular approaches show promise only when considered as a part of plant improvement programs. They are expected to complement, accelerate or enhance conventional breeding efforts. The most successful improvement programs are expected to be those that build on multidisciplinary teams including plant breeders, molecular and cell biologists, plant pathologists, agronomists and animal scientists. The integrated effort of such teams will be critical to the competitive development of marketable forage and turf grass cultivars from molecular breeding programs.

References

A'Brook J, Heard AJ (1975) The effect of ryegrass mosaic virus on the yield of perennial ryegrass swards. Ann Appl Biol 80: 163–168

Allen PG (1987) Insect pests of pasture in perspective. In: Wheeler JL, Pearson CJ, Robards GE (eds) Temperate pastures: their production, use and management. Aust Wool Corp/CSIRO, Canberra, pp 211–225

Akashi R, Adachi T (1992) Plant regeneration from suspension culture-derived protoplasts of apomictic dallisgrass (*Paspalum dilatatum* Poir.). Plant Sci 82: 219–225

Akashi R, Hashimoto A, Adachi T (1993) Plant regeneration from seed-derived embryogenic callus and cell suspension cultures of bahiagrass (*Paspalum notatum*). Plant Sci 90: 73–80

Arsvoll K (1973) Winter damage in Norwegian grasslands 1968–1971. Meld Norg Landbrukshogsk 52: 1–21

Atanassova R, Favet N, Martz F, Chabbert B, Tollier M, Monties B, Fritig B, Legrand M (1995) Altered lignin composition in transgenic tobacco expressing *O*-methyltransferase sequences in sense and antisense orientation. Plant J 8: 465–477

Barnes RF (1990) Importance and problems of tall fescue. In: Kasperbauer MJ (ed) Biotechnology in tall fescue improvement. CRC Press, Boca Raton, pp 2–12

Barnes RF, Baylor JE (1995) Forages in a changing world. In: Barns RF, Miller DA, Nelson CJ (eds) Forages, vol 1. An introduction to grassland Agriculture, 5th edn. Iowa State University Press, Ames, Iowa, pp 3–13

Bashaw EC, Hanna WW (1990) Apomictic reproduction. In: Chapman GP (ed) Reproductive versatility in the grasses. Cambridge University Press, Cambridge, pp 100–130

Bate NJ, Orr J, Ni WT, Meromi A, Nadler-Hassar T, Doerner PW, Dixon RA, Lamb CJ, Elkind Y (1994) Quantitative relationship between phenylalanine ammonia-lyase levels and phenylpropanoid accumulation in transgenic tobacco identifies a rate-determining step in natural product synthesis. Proc Natl Acad Sci USA 91: 7608–7612

Baucher M, Chabbert B, Pilate G, Van Doorsselaere J, Tollier M, Petit-Conil M, Cornu D, Monties B, Van Montagu M, Inze D, Jouanin L, Boerjan W (1996) Red xylem and higher lignin extractability by down-regulating a cinnamyl alcohol dehydrogenase in poplar. Plant Physiol 112: 1479–1490

Beever DE (1993) Ruminant animal production from forages: present position and future opportunities. In: Baker MJ (ed) Grasslands for our world. SIR, Wellington, pp 158–164

Benhamou N, Broglie K, Chet I, Broglie R (1993) Cytology of infection of 35S-bean chitinase transgenic canola plants by *Rhizoctonia solani*: cytochemical aspects of chitin breakdown in vivo. Plant J 4: 295–305

Bernard Vailhé MA, Migné C, Cornu A, Maillot MP, Grenet E, Besle JM, Atanassova R, Martz F, Legrand M (1996a) Effect of modification of *O*-methyltransferase activity on cell wall composition, ultrastructure and degradability of transgenic tobacco. J Sci Food Agric 72: 385–391

Bernard Vailhé MA, Cornu A, Robert D, Maillot MP, Besle JM (1996b) Cell wall degradability of transgenic tobacco stems in relation to their chemical extraction and lignin quality. J Agric Food Chem 44: 1164–1169

Boudet AM, Grima-Pettenati J (1996) Lignin genetic engineering. Mol Breed 2: 25–39

Boudet AM, Goffner DP, Grima-Pettenati J (1996) Lignins and lignification: recent biochemical and biotechnological developments. CR Acad Sci Paris, [Sci Vie /Life Sci] 319: 317–331

Bowman JL, Alvarez J, Weigel D, Meyerowitz EM, Smyth DR (1993) Control of flower development in *Arabidopsis thaliana* by *APETALA1* and interacting genes. Development 119: 721–743

Broglie K, Chet I, Holliday M, Cressman R, Biddle P, Knowlton S, Mauvais CJ, Broglie R (1991) Transgenic plants with enhanced resistance to the fungal pathogen *Rhizoctonia solani*. Science 254: 1194–1196

Buckner RC, Todd JR, Burrus PB, Barnes RF (1967) Chemical composition, palatability, and digestibility of ryegrass-tall fescue hybrids, "Kenwell", and "Kentucky 31" tall fescue varieties. Agron J 59: 345–349

Buxton DR, Russell JR (1988) Lignin constituents and cell-wall digestibility of grass and legume stems. Crop Sci 28: 553–558

Campbell MM, Sederoff RR (1996) Variation in lignin content and composition. Plant Phsiol 110: 3–13

Catherall PL (1966) Effects of barley yellow dwarf virus on the growth and yield of single plants and simulated swards of perennial ryegrass. Ann Appl Biol 57: 155–162

Catherall PL, Parry AL (1987) Effects of barley yellow dwarf virus on some varieties of Italian, hybrid and perennial ryegrasses and their implication for grass breeders. Plant Pathol 36: 148–153

Chapman RA (1979) Disease and nematodes. In: Buckner RC, Bush LP (eds) Tall fescue, no 20 in the series Agronomy. American Society of Agronomy, Crop Science Society of America, Soil Science Society of America, Madison, pp 307–318

Creemers-Molenaar J, Hall RD, Krens FA (1992) Asymmetric protoplast fusion aimed at intraspecific transfer of cytoplasmic male sterility (CMS) in *Lolium perenne* L. Theor Appl Genet 84: 763–770

Dixon RA, Lamb CJ, Masoud S, Sewalt VJH, Paiva NL (1996) Metabolic engineering – prospects for crop improvement through the genetic manipulation of phenylpropanoid biosynthesis and defense responses – a review. Gene 179: 61–71

Dujardin M, Hanna WW (1989) Developing apomictic pearl millet – characterization of a BC3 plant. J Genet Breed 43: 145–151

Dwivedi UN, Campbell WH, Yu J, Datla RSS, Bugos RC, Chiang VL, Podila GK (1994) Modification of lignin biosynthesis in transgenic *Nicotiana* through expression of an antisense *O*-methyltransferase gene from *Populas*. Plant Mol Biol 26: 61–71

Eagling DR, Halloran GM, Sward RJ (1991) Barley yellow dwarf luteovirus in pasture grasses. Rev Plant Pathol 70: 133–143

Eagling DR, Villalta O, Sward RJ (1992) Host range, symptoms and effects on pasture production of a Victorian isolate of ryegrass mosaic virus. Aust J Agric Res 43: 1243–1251

Ealing PM, Hancock KR, White DWR (1994) Expression of the pea albumin 1 gene in transgenic white clover and tobacco. Transgenic Res 3: 344–354

Ebskamp MJM, Van der Meer IM, Spronk BA, Weisbeek PJ, Smeekens SCM (1994) Accumulation of fructose polymers in transgenic tobacco. Bio/Technology 12: 272–275

Ford SA, Baldo BA (1986) A re-examination of ryegrass (*Lolium perenne*) pollen allergens. Int Arch Allergy Appl Immunol 81: 193–202

Forster CA, Wright CE (1970) Variation in the expression of self-fertility in *Lolium perenne* L. Euphytica 19: 61–70

Freidhoff LR, Ehrlich-Kautzky E, Grant JH, Meyer DA, Marsh DG (1986) A study of the human immune response to *Lolium perenne* (rye) pollen and its components, *Lol p*I and *Lol p*II (rye I and rye II). I. Prevalence of reactivity to the allergens among skin test, IgE antibody, and IgG antibody data. J Allergy Clin Immunol 78: 1190–1201

Gan S, Amasino RM (1995) Inhibition of leaf senescence by autoregulated production of cytokinin. Science 270: 1986–1988

Garrett RG, Chu PWG (1997) White clover expressing the coat protein of alfalfa mosaic virus: field trial issues. In: McLean GD, Waterhouse PM, Evans G, Gibbs MJ (eds) Commercialisation of transgenic crops: risks, benefit and trade considerations. Australian Government Publishing Service, Canberra, pp 125–136

Giddings GD, Sackville Hamilton NR, Hayward MD (1997a) The release of genetically modified grasses. Part 1: pollen dispersal to traps in *Lolium perenne*. Theor Appl Genet 94: 1000–1006

Giddings GD, Sackville Hamilton NR, Hayward MD (1997a) The release of genetically modified grasses. Part 2: the influence of wind direction on pollen dispersal. Theor Appl Genet 94: 1007–1014

Goddijn OJM, Pen J (1995) Plants as bioreactors. TIBTECH 13: 379–387

Halpin C, Knight ME, Foxon GA, Campbell MM, Boudet AM, Boon JJ, Chabbert B, Tollier M, Schuch W (1994) Manipulation of lignin quality by downregulation of cinnamyl alcohol dehydrogenase. Plant J 6: 339–350

Hanna WW, Bashaw EC (1987) Apomixis: its identification and use in plant breeding. Crop Sci 27: 1136–1139

Heath R, Huxley H, Stone B, Spangenberg G (1997) cDNA cloning and differential expression of three caffeic acid *O*-methyltransferase homologues from *Lolium perenne*. J Plant Physiol (in press)

Henry M, Dedryver CA (1991) Occurrence of barley yellow dwarf virus in pastures of western France. Plant Pathol 40: 93–99

Hibino T, Takabe K, Kawazu T, Shibata D, Higuchi T (1995) Increase of cinnamaldehyde groups in lignin of transgenic tobacco plants carrying an antisense gene for cinnamyl alcohol dehydrogenase. Biosci Biotech Biochem 59: 929–931

Higgins TJ, O'Brien PA, Spencer D, Schroeder HE, Dove H, Freer M (1989) Potential of transgenic plants for improved amino acid supply for wool growth. In: Rogers GE, Reis PJ, Ward KA, Marshall RC (eds) The biology of wool and hair. Chapman and Hall, London, pp 441–445

Higuchi T, Ito T, Umezawa T, Hibino T, Shibata D (1994) Red-brown color of lignified tissues of transgenic plants with antisense CAD gene: wine-red lignin from coniferyl aldehyde. J Biotech 37: 151–158

Hill KK, Jarvis-Eagan N, Halk EL, Krahn KJ, Liao LW, Mathewson RS, Merlo DJ, Nelson SE, Rashka KE, Loesch-Fries LS (1991) The development of virus-resistant alfalfa, *Medicago Sativa* L. Bio/Technology 9: 373–377

Jenkin TJ (1931) Self-fertility in perennial ryegrass (*Lolium perenne* L.). Bull Welsh Plant Breed Stat Ser H 12: 100–119

Jenkin TJ (1933) Interspecific and intergeneric hybrids in herbage grasses. Initial crosses. J Genet 28: 205–264

Jenkin TJ (1955) Interspecific and intergeneric hybrids in herbage grasses. XV. The breeding affinities of *Festuca rubra*. J Genet 53: 125–130

Jones DG, Heard AJ, Woledge J, Leafe EL, Plumb RT (1977) The effect of ryegrass mosaic virus on carbon assimilation and growth of ryegrasses. Ann Appl Biol 87: 393–405

Jones EL, Roberts E (1991) A note on the relationship between palatability and water-soluble carbohydrates in perennial ryegrass. Aust J Agric Res 30: 163–167

Jung HG, Vogel KP (1986) Influence of lignin on digestibility of forage cell wall material. J Anim Sci 62: 1703–1712

Kajita S, Katayama Y, Omori S (1996) Alterations in the biosynthesis of lignin in transgenic plants with chimeric genes for 4-coumarate:coenzyme A ligase. Plant Cell Physiol 37: 957–965

Khan MRI, Ceriotti A, Tabe L, Aryan A, McNabb W, Moore A, Craig S, Spencer D, Higgins TJV (1996) Accumulation of a sulphur-rich seed albumin from sunflower in the leaves of transgenic subterranean clover (*Trifolium subterraneum* L.). Transgenic Res 5: 179–185

Knox RB, Singh MB, Hough T, Theerakulpisut P (1989) The rye-grass pollen allergen, *Lol p*I. In: El Shami AS, Merrett TG (eds) Advances in the biosciences, Allergy and molecular biology. Pergamon Press, Oxford, pp 161–171

Knox B, Suphioglu C (1996) Environmental and molecular biology of pollen allergens. Trends Plant Sci 1: 156–164

Koltunow AM (1993) Apomixis: embryo sacs and embryos formed without meiosis or fertilization in ovules. Plant Cell 5: 1425–1437

Kortt AA, Caldwell JB, Lilley GG, Higgins TJ (1991) Amino acid and cDNA sequences of a methionine-rich 2S protein from sunflower seed (*Helianthus annuus* L.). Eur J Biochem 195: 329–334

Lagrimini LM, Gingas V, Finger F, Rothstein S, Liu TTY (1997) Characterization of antisense transformed plants deficient in the tobacco anionic peroxidase. Plant Physiol 114: 1187–1196

Leblanc O, Armstead I, Pessino S, Ortiz JPA, Evans C, do Valle C, Hayward MD (1997) Non-radioactive mRNA fingerprinting to visualise gene expression in mature ovaries of *Brachiaria* hybrids derived from *B. brizantha*, an apomictic tropical forage. Plant Sci (in press)

Legris G (1996) Asymmetric somatic hybridization between red fescue (*Festuca rubra*) and perennial ryegrass (*Lolium perenne*). PhD Diss, Swiss Federal Institute of Technology, Zurich

Li XM, Nield J, Hayman D, Langridge P (1994) Cloning a putative self-incompatibility gene from the pollen of the grass *Phalaris coerulescens*. Plant Cell 6: 1923–1932

Mandel MA, Yanofsky MF (1995) A gene triggering flower formation in *Arabidopsis*. Nature 377: 522–524

Marsh DG (1975) Allergens and the genetics of allergy. In: Sela A (ed) The antigens. Academic Press, New York, pp 271–350
Masoud, SA, Zhu Q, Lamb C, Dixon RA (1996) Constitutive expression of an inducible β-1,3-glucanase in alfalfa reduces disease severity caused by the oomycete pathogen *Phytophthora megasperma* f. sp. *medicaginis*, but does not reduce disease severity of chitin-containing fungi. Transgenic Res 5: 313–323
Matsumoto N, Araki T (1982) Field observation of snow mould pathogens of grasses under snow cover in Sapporo. Res Bull Hokkaido Nat Agric Exp Stn 135: 1–10
McBride KE, Svab Z, Schaaf DJ, Hogan PS, Stalker DM, Maliga P (1995) Amplification of a chimeric *Bacillus* gene in chloroplasts leads to an extraordinary level of an insecticidal protein in tobacco. Bio/Technology 13: 362–365
McIntyre CL, Bettenay HM, Manners JM (1996) Strategies for the suppression of peroxidase gene expression in tobacco. II. In vivo suppression of peroxidase activity in transgenic tobacco using ribozyme and antisense constructs. Transgenic Res 5: 263–270
McKersie BD (1997) Improving forage production systems using biotechnology. In: McKersie BD, Brown DCW (eds) Biotechnology and the improvement of forage legumes. CAB International, Wallingford, pp 3–21
Mett VL, Lochhead LP, Reynolds PHS (1993) Copper-controllable gene expression system for whole plants. Proc Natl Acad Sci USA 90: 4567–4571
Mett VL, Podivinsky E, Tennant AM, Lochhead LP, Jones WT, Reynolds PHS (1996) A system for tissue-specific copper-controllable gene expression in transgenic plants: nodule-specific antisense of asparate aminotransferase-P_2. Transgenic Res 5: 105–113
Meyer WA (1982) Breeding disease resistant cool-season turfgrass cultivars for the United States. Plant Dis 66: 341–344
Meyer WA, Funk CR (1989) Progress and benefits to humanity from breeding cool-season grasses for turf. In: Sleper DA, Asay KH, Pedersen JF (eds) Contributions from breeding forage and turf grasses. CSSA Spec Publ 15: 31–48
Michell PJ (1973) Relations between fibre and water soluble carbohydrate contents of pasture species and their digestibility and voluntary intake by sheep. Aust J Exp Agric Anim Husb 13: 165–170
Ni WT, Paiva NL, Dixon RA (1994) Reduced lignin in transgenic plants containing a caffeic acid O-methyltransferase antisense gene. Transgenic Res 3: 120–126
Ong EK, Griffith IJ, Knox RB, Singh MB (1993) Cloning of a cDNA encoding a group-V (group-IX) allergen isoform from ryegrass pollen that demonstrates specific antigenic immunoreactivity. Gene 134: 235–240
Perez M, Ishioka GY, Walker LE, Chesnut RW (1990) cDNA cloning and immunological characterization of the ryegrass allergen *Lol p*I. J Biol Chem 265: 16210–16215
Pessino SC, Ortiz JPA, Leblanc O, do Valle CB, Evans C, Hayward MD (1997) Identification of a maize linkage group related to apomixis in *Brachiaria*. Theor Appl Genet 94: 439–444
Pilon-Smits EAH, Ebskamp MJM, Paul MJ, Jeuken MJW, Weisbeek PJ, Smeekens SCM (1995) Improved performance of transgenic fructan-accumulating tobacco under drought stress. Plant Physiol 107: 125–130
Radojevic I, Simpson RJ, St John JA, Humphreys MO (1994) Chemical composition and in vitro digestibility of lines of *Lolium perenne* selected for high concentrations of water-soluble carbohydrate. Aust J Agric Res 45: 901–912
Reed KFM (1994) Improved grass cultivars increase milk and meat production – a review. N Z J Agric Res 37: 277–286
Reis PJ (1988) The influence of absorbed nutrients on wool growth. In: Rogers GE, Reis PJ, Ward KA, Marshall RC (eds) The biology of wool and hair. Chapman and Hall, London, pp 185–203
Reis PJ, Schinckel PG (1963) Some effects of sulfur-containing amino acids on the growth and composition of wool. Aust J Biol Sci 16: 218–230
Rogers GE (1990) Improvement of wool production through genetic engineering. TIBTECH 8: 6–11

Schroeder HE, Khan MRI, Knibb WR, Spencer D, Higgins TJV (1991) Expression of a chicken ovalbumin gene in three lucerne cultivars. Aust J Plant Physiol 18: 495–505

Schumann GL, Wilkinson HT (1992) Research methods and approaches to the study of diseases in turfgrasses. In Waddington DV, Carrow RN, Shearman RC (eds) Turfgrass, no 32 in the series Agronomy. American Society of Agronomy, Crop Science Society of America, Soil Science Society of America, Madison

Sewalt VJH, Ni W, Jung HG, Dixon RA (1997a) Lignin impact on fiber degradation: increased enzymatic digestibility of genetically engineered tobacco (*Nicotiana tabacum*) stems reduced in lignin content. J Agric Food Chem 45: 1977–1983

Sewalt VJH, Ni W, Blount JW, Jung HG, Masoud SA, Howles PA, Lamb C, Dixon RA (1997b) Reduced lignin content and altered lignin composition in transgenic tobacco down-regulated in expression of L-phenylalanine ammonia-lyase or cinnamate 4-hydrolase. Plant Physiol 115: 41–50

Sidoli A, Tamborini E, Giuntini I, Levi S, Volonté G, Paini C, De Lalla C, Siccardi AG, Baralle FE, Galliani S, Arosio P (1993) Cloning, expression, and immunological characterization of recombinant *Lolium perenne* allergen *Lol p*II. J Biol Chem 268: 21819–21825

Singh MB, Hough T, Theerakulpisut P, Avjioglu A, Davies S, Smith PM, Taylor P, Simpson RJ, Ward LD, McCluskey J, Puy R, Knox RB (1991) Isolation of cDNA encoding a newly identified major allergenic protein of rye-grass pollen: intracellular targeting to the amyloplast. Proc Natl Acad Sci USA 88: 1384–1388

Smart IJ , Tuddenham WG, Knox RB (1979) Aerobiology of grass pollen in the city atmosphere of Melbourne: effects of weather parameters and pollen sources. Aust J Bot 27: 333–342

Smith JD, Jackson N, Woolhouse AR (1989) Fungal diseases of amenity turf grasses. E & FN Spon, London

Smith KF, Reed KFM, Foot JZ (1997) An assessment of the relative importance of specific traits for the genetic improvement of nutritive value in dairy pasture. Grass Forage Sci 52: 167–175

Spangenberg G, Wang ZY, Wu XL, Nagel J, Potrykus I (1995) Transgenic perennial ryegrass (*Lolium perenne*) plants from microprojectile bombardment of embryogenic suspension cells. Plant Sci 108: 209–217

Spangenberg G, Wang ZY, Heath R, Kaul V, Garrett R (1997) Biotechnology in pasture plant improvement: methods and prospects. Proc XVIII Int Grassland Congr, Winnipeg and Saskatoon, June 8–19, vol 3

Sprenger N, Bortlik K, Brandt A, Boller T, Wiemken A (1995) Purification, cloning, and functional expression of sucrose:fructan 6-fructosyltransferase, a key enzyme of fructan synthesis in barley. Proc Natl Acad Sci USA 92: 11652–11656

Stewart D, Yahiaoui N, McDougall GJ, Myton K, Marque C, Boudet AM, Haigh J (1997) Fourier-transform infrared and Raman spectroscopic evidence for the incorporation of cinnamaldehydes into the lignin of transgenic tobacco (*Nicotiana tabacum* L.) plants with reduced expression of cinnamyl alcohol dehydrogenase. Planta 201: 311–318

Stone BA (1994) Propects for improving the nutritive value of temperate, perennial pasture grasses. N Z J Agric Res 37: 349–363

Strizhov N, Keller M, Mathur J, Konczkalman Z, Bosch D, Prudovsky E, Schell J, Sneh B, Koncz C, Zilberstein A (1996) A synthetic cryIC gene, encoding a *Bacillus thuringiensis* delta-endotoxin, confers *Spodoptera* resistance in alfalfa and tobacco. Proc Natl Acad Sci USA 93: 15012–15017

Tabe LM, Wardley-Richardson T, Ceriotti A, Aryan A, McNabb W, Moore A, Higgins TJV (1995) A biotechnological approach to improving the nutritive value of alfalfa. J Anim Sci 73: 2752–2759

Tamborini E, Brandazza A, De Lalla C, Musco G, Siccardi AG, Arosio P, Sidoli A (1995) Recombinant allergen *Lol p II*: expression, purification and characterization. Mol Immunol 32: 505–513

Thorogood D, Hayward MD (1988) The genetic control of self-fertility in *Lolium temulentum.* Heredity 61: 288

Thorogood D, Hayward MD (1991) The genetic control of self-incompatibility in an inbred line of *L. perenne*. Heredity 67: 175–181
Thorogood D, Hayward MD (1992) Self-compatibility in *Lolium temulentum* L.: its genetic control and transfer into *L. perenne* L. and *L. multiflorum* Lam. Heredity 68: 71–78
Van der Meer IM, Ebskamp MJM, Visser RGF, Weisbeek PJ, Smeekens SCM (1994) Fructan as a new carbohydrate sink in transgenic potato plants. Plant Cell 6: 561–570
Van Doorsselaere J, Baucher M, Chognot E, Chabbert B, Tollier M, Petit-Conil M, Leple J, Pilate G, Cornu D, Monties B, Van Montagu M, Inze D, Boerjan W, Jouanin L (1995) A novel lignin in poplar trees with a reduced caffeic acid/5-hydroxyferulic acid *O*-methyltransferase activity. Plant J 8: 855–864
Van Wijk AJP, Boonman JG, Rumball W (1993) Achievements and prospectives in the breeding of forage grasses and legumes. In: Baker MJ (ed) Grasslands for our world. SIR, Wellington, pp 116–120
Voisey CR, White DWR, Wigley PJ, Chilcott CN, Mcgregor PG, Woodfield DR (1994) Release of transgenic white clover plants expressing *Bacillus thuringiensis* genes – an ecological perspective. Biocontrol Sci Technol 4: 475–481
Wandelt CI, Khan MRI, Craig S, Schroeder HE, Spencer D, Higgins TJV (1992) Vicillin with carboxy-terminal KDEL is retained in the endoplasmic reticulum and accumulates to high levels in the leaves of transgenic plants. Plant J 2: 181–192
Weigel D, Nilsson O (1995) A developmental switch sufficient for flower initiation in diverse plants. Nature 377: 495–500
Whetten R, Sederoff R (1995) Lignin biosynthesis. Plant Cell 7: 1001–1013
Wilkins PW, Catherall PL (1977) Variation in reaction to barley yellow dwarf virus in ryegrass and its inheritance. Ann Appl Biol 85: 257–263
Winter W (1986) Auswinterungsschäden in Raigras-Reinbeständen, die Bedeutung pilzlicher Krankheitserreger und ihre mögliche Eindämmung, I. Ergebnisse aus den Feldversuchen mit natürlicher Infektion. J Phytopathol 117: 226–243
Wu XL, Ye XD, Wang ZY, Potrykus I, Spangenberg G (1997) Gene transfer to ryegrasses: down-regulation of major pollen allergens in transgenic plants. Proc XVIII Int Grassland Congr, Winnipeg and Saskatoon, June 8–19, vol 3
Ye XD (1997) Gene transfer to ryegrasses (*Lolium* spp.): modification of fructan metabolism in transgenic plants. PhD Diss, Swiss Federal Institute of Technology, Zurich
Ye X, Wang ZY, Wu XL, Potrykus I, Spangenberg G (1997) Transgenic Italian ryegrass (*Lolium multiflorum*) plants from microprojectile bombardment of embryogenic suspension cells. Plant Cell Rep 16: 379–384
Zhu Q, Maher EA, Masoud S, Dixon RA, Lamb CJ (1994) Enhanced protection against fungal attack by constitutive co-expression of chitinase and glucanase genes in transgenic tobacco. Bio/Technology 12: 807–812

Subject Index

The manufacturer's authorised representative in the EU is Springer Nature Customer Service Centre GmbH, Europaplatz 3, 69115 Heidelberg, Germany. If you have any concerns regarding our products, please contact ProductSafety@springernature.com

Printed and bound by CPI Group (UK) Ltd, Croydon, CR0 4YY
22/09/2026
02221059-0004